T0413541

Nanofibers of Conjugated Polymers

Nanofibers of Conjugated Polymers

A. Sezai Sarac

Pan Stanford Publishing

Published by

Pan Stanford Publishing Pte. Ltd.
Penthouse Level, Suntec Tower 3
8 Temasek Boulevard
Singapore 038988

Email: editorial@panstanford.com
Web: www.panstanford.com

British Library Cataloguing-in-Publication Data
A catalogue record for this book is available from the British Library.

Nanofibers of Conjugated Polymers

ISBN 978-981-4613-51-4 (Hardcover)
ISBN 978-981-4613-52-1 (eBook)

Printed in the USA

Contents

Preface xi

1. Introduction **1**
1.1 Introduction 1
1.1.1 Polymeric Nanofibers 1
1.1.2 Nanofiber Production with Other Techniques 6
1.2 Previous Development 7
1.3 Electrospinning Setup 8
1.4 Uses and Types of Electrospinning 10
1.5 Parameters Affecting Electrospinning 11
1.5.1 Polymer Solution Parameters 12
1.5.1.1 Surface tension 12
1.5.1.2 Solution conductivity 16
1.5.1.3 Dielectric effect 16
1.5.1.4 Solution viscosity 20
1.5.1.5 Volatility of the solvent 22
1.5.2 Polymer Processing Parameters 23
1.5.2.1 Applied voltage 23
1.5.2.2 Flow rate 23
1.5.2.3 Distance 24
1.5.2.4 Effect of the collector 24
1.5.2.5 Diameter of the needle 25
1.6 Theoretical Considerations in Electrospinning 25
1.6.1 Electrospinning of Solutions in the Semidilute Concentration Regime 30
1.7 Effect of Polymer Concentration and Filler 34
1.8 Conductive Nanofibrils by Coating Cellulose 42

2. Applications of Nanofibers **45**
2.1 Applications of Nanofibers 45
2.1.1 Filtration Applications 45
2.1.2 Nanocomposites 46
2.1.3 Biomedical Applications 53
2.1.4 Agricultural, Electrical, Optical, and Other Applications 61

2.2 Electrospun Ceramic Nanofibers and Their Applications 63
2.3 Electrospun Metallic Nanofibers 65
2.4 Electrospun Carbon/Graphite Nanofibers 65

3. Conjugated Polymers **71**
3.1 Conjugated Polymers 71
3.1.1 Conductivity of Conjugated Polymers 71
3.1.2 Electronic Conduction 72
3.2 The Molecular Structure of PEDOT 74
3.2.1 Doping in Poly(*p*-Phenylene Vinylene) and Poly(3,4-Ethylenedioxythiophene) 75
3.2.2 Chemical Doping: Protonation of Polyaniline 85
3.3 Applications of Conducting Polymers 88
3.3.1 Electro-Optic Activity 89
3.3.1.1 Excitons and free charges 89
3.3.2 Tissue Engineering and Sensor Applications 91
3.4 Alkyl-Substituted Polythiophenes 93
3.4.1 Regioregularity 94
3.4.2 Solubility 95
3.4.3 Chemical Oxidative Polymerization 96
3.4.4 Catalyst for the Deprotonation of Oligomers of 3-Methylthiophene 99
3.5 Poly(3,4-Ethylenedioxythiophene) Composites 100
3.5.1 Poly(3,4-Ethylenedioxythiophene)/ Polystyrene Sulfonic Acid 101
3.5.2 Conductivity Enhancement of PEDOT-PSS 102
3.5.3 Charge Transport in the Conducting Polymer PEDOT-PSS 102
3.5.4 Morphology of PEDOT-PSS 102
3.6 Copolymers of PEDOT 103
3.7 Supercapacitors 106
3.8 Electrochemical Impedance Spectroscopy 108

4. Polymerization Techniques **111**
4.1 Polymerization Techniques 111

4.2 Synthesis and Characterization of PEDOT and PEDOT/PSS 112
4.2.1 Oxidative Chemical Polymerization of EDOT-Based Monomers 112
4.3 PEDOT as an Electrode Material for Solid Electrolyte Capacitors 113
4.3.1 Preparation of Poly(Vinyl Acetate)/Poly(3,4-Ethylenedioxythiophene) Poly(Styrene Sulfonate) Composites 114
4.3.2 Synthesis of Poly(3,4-Ethylenedioxythiophene) in a Poly(Vinyl Acetate) Matrix 115
4.3.3 Preparation of Electrospinning Solutions: PEDOT in PVAc Matrix 116
4.3.3.1 Electrospinning of PEDOT-PSS/PVAc 116
4.4 Conjugated Polymeric Nanostructures 118
4.4.1 Electrochemical Polymerization of Heterocyclic and Aromatic Monomers 120
4.4.2 Composites with Carbon Nanomaterials 122
4.4.3 Composites with Insulating Polymers 123
4.4.4 Composites with Metal or Oxide Nanoparticles 124
4.4.5 Applications of Electrosynthesized CP Nanomaterials 125
4.4.5.1 Sensors 125
4.4.5.2 Fuel cell electrode 126
4.4.5.3 Batteries 127
4.4.5.4 Electrochemical actuators 127
4.5 MWCNT/PS and PPy/PS Nanofibers 128

5. Electrospinning: The Velocity Profile 131
5.1 Electrospinning: The Velocity Profile 131
5.2 Electrospinning and Nanofibers: Applications 135
5.3 Electrospinning of Polyaniline Blends 138
5.3.1 Effect of Rotating Speed on Fiber Alignment 143

5.3.1.1 Yarn formation 144
5.3.2 Conductivity and Mechanical Properties of PEDOT Composite Fibers 146
5.4 Characterization of PEDOT-PSS/PVAc Composites 146
5.4.1 FTIR-ATR Spectrophotometric Analysis 146
5.4.2 UV-Vis Spectrophotometric Analysis 149
5.4.3 Morphological Analysis 151
5.4.4 Dynamic Mechanical Analysis 153
5.4.5 BET Surface Area 155
5.4.6 Broadband Dielectric Spectrometer 156
5.4.7 Organic Vapor-Sensing Characteristics of PEDOT-PSS/PVP and PVP Nanofibers 158

6. Impedance Spectroscopy and Spectroscopy on Polymeric Nanofibers 161
6.1 Impedance Spectroscopy on Polymeric Nanofibers 161
6.2 Impedance Spectroscopy: A General Overview 162
6.2.1 Equivalent Circuits Modeling 162
6.3 Modeling Charge Transport within the Conjugated Polymer Film 164
6.4 Synthesis, Characterization, and Electrochemical Impedance Spectroscopy of PEDOT-PSS/PVAc and PEDOT/PVAc 168
6.5 Characterization of Synthesized PEDOT in PVAc Matrix by FTIR-ATR, UV-Vis Spectrophotometric Analysis 171
6.6 Electrochemical Impedance Spectroscopy of Nanofiber Mats on ITO-PET 173
6.7 Molecular Structure and Processibility of Polyanilines 175
6.8 Solubility and Processability 177
6.9 Substitution onto the Backbone 181
6.9.1 Sulfonated Polyaniline 181

6.9.2 Poly(Anthranilic Acid) 182
6.9.3 Preparation of Poly(Anthranilic Acid)/Polyacrylonitrile Blends and Electrospinning Parameters 184
6.9.4 UV-Vis Spectrophotometeric Investigation of Polymer Solutions 184
6.9.5 FTIR-ATR Spectrophotometric Analysis of PAN/PANA Nanofibers 187
6.9.6 Morphology of Electrospun Nanofibers of PANA/PAN Blends 189
6.9.7 Cyclic Voltammetry of Nanofibers of PANA/PAN Blends 190
6.9.8 Dynamic Mechanical Analysis 191
6.9.9 Electrochemical Impedance Spectroscopy 191

7. Electrochemical Capacitive Behavior of Nanostructured Conjugated Polymers 197

7.1 Electrochemical Capacitive Behavior of Nanostructured Conjugated Polymers 197
7.1.1 PANI Electrochemical Supercapacitor 198
7.2 Polypyrrole and Its Composites 201
7.2.1 Synthesis of Polypyrrole and PEDOT Carbon Nanofiber Composites 203
7.2.2 Synthesis and Pseudocapacitance of Chemically Prepared PPy 206
7.2.3 Graphite/Polypyrrole Composite Electrode 208
7.3 EIS Study of Poly(3,4-Ethylenedioxythiophene) 210
7.3.1 Carbon Nanofibers and PEDOT-PSS Bilayer Systems 212

8. Preparation of Conductive Nanofibers 215

8.1 Preparation of Conductive Nanofibers 215
8.2 Polyaniline Nanofibers 216
8.3 Addition of Nanoparticles 222
8.4 Polypyrrole Nanofibers with Carriers 223
8.5 Polyurethane/Polypyrrole Composite Nanofibers 229
8.6 Doped Nanofibers and Conductivities 238

8.7 Light-Emitting Polymer Nanofibers 244
8.7.1 Light Emission and Waveguiding in Conjugated Polymer Nanofibers Electrospun from Organic Salt–Added Solutions 245
8.8 Bioapplications 247
8.8.1 Polymer Nanofibers for Biomedical, Biotechnological and Capacitor Applications 250
8.9 Concluding Remarks 252

Bibliography **255**

Index 277

Preface

Conjugated polymer composites with high dielectric constants are being developed by the electronics industry in response to the need for power-grounded decoupling to secure the integrity of high-speed signals and to reduce electromagnetic interference.

Electrically conducting polymers are materials which simultaneously possess the physical and chemical properties of organic polymers and the electronic characteristics of metals. Electrospinning based on the application of a static electric field on a polymer solution or melt through a spinneret appears to be a simple and well-controllable technique able to produce polymeric nanofibers. It is a versatile method for generating ultrathin fibers from a rich variety of materials that include polymers, composites, and ceramics. Due to its good adhesion to a number of substrates, and to some extent because it can be produced in large quantities, it can be used in emulsions, paints, adhesives, and various textile-finishing operations.

Conductive materials in fibrillar shape may be advantageous compared to films due to their inherent properties such as anisotropy, high surface area, and mechanical strength. Fibrous conductive materials are of particular interest in electroactive composites. Fine metal nanoparticles, carbon fibers, and carbon nanotubes have been efficiently distributed in an insulating polymer matrix in order to improve both electrical and mechanical properties.

Combination of electrical properties with good mechanical performance is of particular interest in electroactive polymeric technology. Fibers have intrinsically high structure factor that results in lower percolation threshold values avoiding material fracture with low filler content. Also, the use of mechanically stronger fibers will result in stronger composites.

Multifunctional micro- and nanostructures of conjugated polymers have received great attention. Some of them have several advantages, for example, pyrrole, aniline, and 3,4-etylenedioxythiopene have properties of easy polymerization, high conductivity, and good thermal stability; but disadvantages of brittleness and hard processibility can be overcome by the production of their nanocomposites. Conjugated polymer composites as a nanofiber mat with different dielectric properties can be used in electronics industry, sensors, batteries, and electrical stimulation to enhance the nerve-regeneration process and for the construction of scaffolds for nerve-tissue engineering. Electrospinning is a technique used for the production of thin continuous fibers from a variety of materials including blends and composites. The extremely small diameters (~ nm) and high surface-to-volume and aspect ratios found in electrospun fibers cannot be achieved through conventional spinning.

This book covers general aspects, fundamental concepts, and equations of electrospinning used for the production of nanofibers and reviews latest researches on inclusion of conjugated polymer in different polymeric structures such as composites or blends of conjugated polymer nanofibers obtained by electrospinning.

Prof. Dr. A. Sezai Sarac
Summer 2016

Chapter 1

Introduction

1.1 Introduction

1.1.1 Polymeric Nanofibers

Unlike conventional fiber spinning techniques, wet spinning, dry spinning, melt spinning, and gel spinning, which are capable of producing polymer fibers with diameters down to the micrometer range, "electrospinning" is capable of producing polymer fibers in the nanometer diameter range. Electrospinning is a novel and efficient fabrication process that can be utilised to assemble fibrous polymer mats composed of fiber diameters ranging from several microns down to fibers with diameter lower than 100 nm. This method uses electrically charged jet of polymers or liquid states of polymers in order to make fibers from microdimensions to nanodimensions by using a high-voltage electric field to form solid fibers from a polymeric fluid stream (solution or melt) delivered through a millimeter-scale nozzle.[1]

Polymer nanofibers exhibit several properties making them favorable for different applications. Nanofibers have a high specific surface area due to their small diameters, and nanofiber mats can

Nanofibers of Conjugated Polymers
A. Sezai Sarac

ISBN 978-981-4613-51-4 (Hardcover), 978-981-4613-52-1 (eBook)
www.panstanford.com

be highly porous with good pore interconnection. Electrospinning is a simple and versatile method for generating ultrathin fibers from a rich variety of materials that include polymers, composites, and ceramics, that is, poly(vinyl acetate) (PVAc) is used as a carrier polymer for preparation of conductive or inorganic nanofibers.[1]

Polymer nanofibers can be obtained by applying electrical force at the surface of a polymer solution. A charged jet is ejected to the tip of the needle, and the jet extends, bends, and then follows a looping and spiraling path due to the action of the electrical field. It becomes very thin, until it reaches the collector. Nanofibers that have diameters from several nanometers to hundreds of nanometers can be obtained in the form of nonwoven fiber mats. The small diameters lead to a large surface-area-to-mass ratio, a porous structure with excellent pore interconnectivity, and extremely small pore dimensions.[1–3]

Since an electrospinning apparatus includes a syringe, a syringe pump, a spinneret, a collector, and a high-voltage power supply, there are mainly two types of parameters, system parameters and process parameters. Viscosity, concentration, surface tension, molecular weight, conductivity, and dielectric of the polymer solution are system parameters. Applied voltage, feeding rate, tip-to-collector distance, heat of the solution, and ambient parameters are process parameters. The fiber morphology depends on the polymer type, conformation of polymer chain, system and process parameters.[1,4,5]

These thin fibers have been used in filtration, protective clothing, tissue engineering scaffolds, sensors, energy storage, battery separators, composite materials, and biomedical applications such as wound dressing and drug delivery. Fundamental requirements for all cases are controlled pore sizes, an enhanced specific surface area, and permeation properties.[6] There are detailed reports on the electrospinning method and its applications.[7–10] Many applications of electrospun nanofibers, such as filtration, affinity membranes, tissue engineering scaffolds, wound healing, release control, catalyst and enzyme carriers, sensors, energy conversion, and storage, were already reported. Applications of electrospun nanofibers in electronics, such as electrodes, separators, electrolytes, sensors, and actuators, controlled structures, morphologies, and mechanical properties of electrospun nanofibers were evaluated.[11] Applications of electrospinning for energy conversion and storage devices such as solar cells, fuel cells, lithium ion batteries, and supercapacitors were studied.[12]

In fuel cells, well known catalyst is produced from carbon black–supported Pt particles (Pt/C) for hydrogen and oxygen redox reactions which occurs at anode and cathode; but conventional Pt/C catalyst has low durability and can be easily poisoned by carbon monoxide. Electrospun Pt/ruthenium, Pt/rhodium, and Pt nanowires have been produced and compared with Pt/C showing better performance in a proton exchange membrane fuel cell (PEMFC).

The combination of conventional Pt/C on electrospun Pt nanowires was also produced, indicating improved oxygen reduction reaction activity. Addionally, electrospun polyaniline (PANI) and carbon nanofibers had been used as a catalyst support in direct methanol fuel cells (DMFCs) for improvement of methanol oxidation reaction. Conductive materials in fibrillar form can be advantageous compared to films due to their anisotropic behaviour, high surface area, and mechanical strength. Conductivematerials in fiber form are of specific interest in electroactive composites. Very small metal nanoparticles, carbon fibers, and carbon nanotubes have been well distributed in an insulating polymer matrix by improving electrical and mechanical properties of material. The combination of electrical properties with good mechanical properties is important in electroactive polymeric technology.

The combination of electrical properties with good mechanical performance is of particular interest in electroactive polymeric technology. Fibers have an intrinsically high structure factor, which results in lower percolation threshold values, avoiding material fracture with low filler content. Also, the use of mechanically stronger fibers will result in stronger composites.

The electrical performance of conjugated fiber composites is investigated in the gelation process in a polymer after crosslinking, where the conductive network corresponds to the gel fraction. If a fiber segment is able to conduct it must be connected to the gel in both ends.

In the case of low filler concentration, there is no conductive network; if the fiber content is increased until a three-dimensional network is formed, percolation is achieved and conductivity increases. And when some processing technique is able to spatially orient the fibers, the conductive network will be formed at even lower concentrations, and the electrically conductive polymers will have good electrical performance associated to the mechanical properties of the matrix.

There are several techniques that allow to produce polymeric nanofibers, such as drawing, template synthesis, phase separation, self-assembly, solution blow spinning, and electrospinning. The drawing process requires a polymer with appropriate viscoelastic properties that is able to be deformed and kept connected by cohesive forces. Besides being simple and inexpensive, this technique is very limited for conjugate polymers, since most of them have lower solubility and form solutions with a small viscous modulus.

In the phase separation approach, a polymer is solubilized and then undergoes the gelation process. Due to the physical incompatibility of the gel and the solvent, solvent is removed and the remaining structure, after freezing, is obtained in nanofibrilar form. Template synthesis implies the use of a template or mold to obtain a desired structure. Commonly metal oxide membranes with nanopores are used, where a polymer solution is forced to pass through to a nonsolvent bath, originating nanofibers, depending on the pores' diameter.

A similar method was developed to obtain polyaniline (PANI) nanofibrils in a way that growing polymer chains separate from the solution according to their molecular size.

Polyaniline nanofibers are observed to be formed spontaneously during the chemical oxidative polymerization of aniline. The key of the nanofibril formation is the suppression of the secondary chain growth that leads to agglomerated particles. Depending on the doping acid, nanofibers with diameters values up to ~100 nm can be obtained by this approach. These nanofibrils can be used as a template to grow inorganic/PANI nanocomposites that might lead to electrical bistability that can be used for nonvolatile memory devices.

Electrospinning based on the application of a static electric field on a polymer solution or melt through a spinneret appears to be a simple and well-controllable technique able to produce polymeric nanofibers. A typical experimental setup is based on a capillary injection tip, a high-voltage source able to apply electric fields of 100–500 KVm^{-1}, and a metallic collector, or counter electrode. Electric current in electrospinning experiments is usually in the order of a few milliamperes.[13]

An electrospinning unit is a system for producing ultrafine fibers with a diameter of about 20–1000 nm. A nanofiber has a very high specific surface area, a small diameter, and large porosity. There are about hundred kinds of polymers that could be used as raw materials.

Electrospinning is applicable to a wide range of polymers like those used in conventional spinning, that is, polyolefine, polyamides, polyester, aramide, and acrylic, as well as biopolymers like proteins, DNA, and polypeptides, or others like electrically conducting, photonic and other polymers such as poly(ethylene oxide) (PEO), DNA, poly(acrylic acid) (PAA), poly(lactic acid) (PLA), and also collagen, organics such as nylon, polyester, and acryl resin, and poly(vinyl alcohol) (PVA), polystyrene (PS), polyacrylonitrile (PAN), peptide, cellulose, etc.

The high specific surface area and small pore size of electrospun nanofibers make them good candidates for a wide variety of applications. For instance, nanofibers with a diameter of ~100 nm have a ratio of geometrical surface area to mass of ~100 m^2/g.[14] Another aspect of using electrospun fibers is that the fibers may dissipate or retain the electrostatic charges, depending on the electrical properties of the polymer. The charges can be manipulated as well by electrical fields, and the electrical polarity of the fibers is affected by the polarity of the applied voltage.[15]

Electrospinning of various polymers and biopolymers are a candidate for certain applications, including multifunctional membranes, wound dressing, drug delivery, and biomedical structural elements. Scaffolds used in tissue engineering, artificial organs, and vascular grafts are some examples of tissue engineering applications. Protective shields in speciality fabrics, filter media for submicron particles in the separation industry, composite reinforcement, and structures for nanoelectronic machines are among others.[9]

The process makes use of electrostatic and mechanical forces to spin fibers from the tip of a fine spinneret. The spinneret is maintained at a positive or a negative charge by a direct current (DC) power supply. When the electrostatic repelling force overcomes the surface tension force of the polymer solution, the liquid spills out of the spinneret and forms an extremely fine continuous filament.[16]

1.1.2 Nanofiber Production with Other Techniques

Generally, fibers can be defined as objects or materials that have an elongated structure, as shown in Fig. 1.1. There are other definitions according to the field they are used in, such as the textile industry, biochemistry, botany, physiology, and anatomy. With regard to fibers, "nano" refers to the diameter of the fiber. However, fibers as less than 1 μm are accepted as nanofibers. In the industry, it is acceptable to classify fibers up to 500 nm as "nano," whereas some scientists use the term "submicron."

Nanofibers present a high surface-area-to-volume ratio, better mechanical properties, for example, good directional strength, and flexibility, so they can be utilized for a wide variety of materials and applications, including for their mechanical, biomedical, optical, electronical, and chemical properties.

A comparison between different techniques is given in Table 1.1. Among all, electrospinning is the best candidate for further development, with a wide range of opportunities to be utilized in all types of polymers (both synthetic and natural) and ceramics.

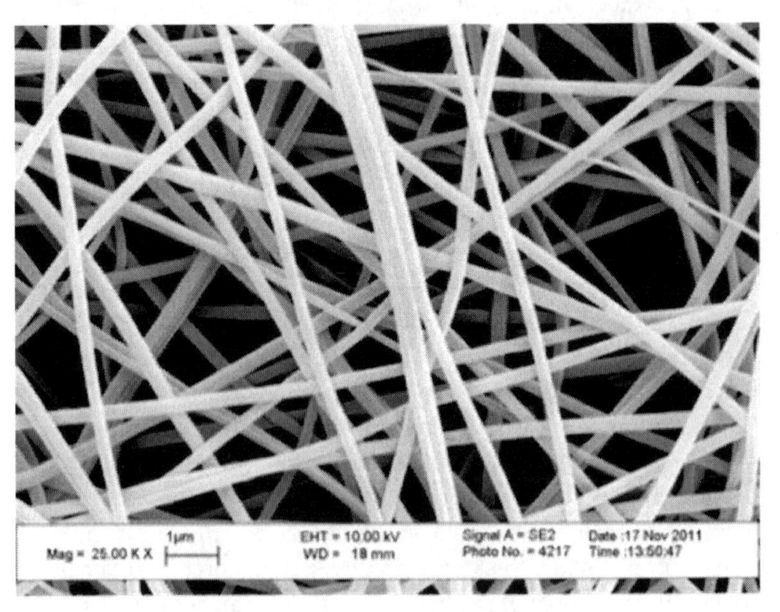

Figure 1.1 SEM photograph of a poly(*N*-vinyl pyrrolidone) (PVP) nanofiber (unpublished result of A. Sezai Sarac).

Table 1.1 Electrospinning compared to other related techniques

Process	Advantages	Disadvantages
Drawing	It requires simple equipment.	No continuous fibers, no control on fiber dimension
Template Synthesis	Fibers of different diameters can be obtained by using different templates.	Process not easily scaled up
Phase Separation	The process can directly fabricate a nanofiber matrix. Mechanical properties of the matrix can be arranged by polymer concentration.	For specific polymers
Self-Assembly	It is good for smaller (10 nm) nanofibers.	No control on fiber diameter and complexity of the process
Electrospinning	It is cost-effective and gives long, continuous nanofibers.	Jet instability; control on fiber diameter

1.2 Previous Development

For fiber production using electrostatic forces, electrostatic spinning relies on an electrically charged jet of polymers or liquid states of polymers in order to make fibers from microdimensions to nanodimensions. In contrast to fibers created from conventional melt spinning, dry spinning, or wet spinning, they possess several good properties. Electrospun fibers are smaller in diameter and longer in length so that they have very high surface-area-to-volume ratios and fibers are placed closer to each other on the mat when compared to fibers produced from dry- or wet-spinning technologies.

In late 1800s Rayleigh investigated the hydrodynamic stability of a liquid jet under electrical field, he showed that if the electrostatic force overcomes the surface tension—acting as opposite direction of the electrostatic force—liquid is thrown out in fine jets.

Although the term "electrospinning," derived from "electrostatic spinning," was used in ~1994, its fundamental idea dates back more than 65 years earlier. From 1934 to 1944, patents were published describing an experimental setup for the production of polymer filaments using an electrostatic force. A polymer solution, such as cellulose acetate, was introduced into the electric field. The polymer filaments were formed, from the solution, between two electrodes bearing electrical charges of opposite polarity. One of the electrodes was placed into the solution and the other onto a collector. Once ejected out of a metal spinnerette with a small hole, the charged solution jets evaporated to become fibers that were collected on the collector. The main difference depended on the properties of the spinning solution, such as polymer molecular weight and viscosity. If the distance between the spinnerette and the collecting device was small, the fibers tended to stick to the collecting device and also to each other due to incomplete solvent evaporation.[9]

Ultrafine fibers or fibrous structures of different polymers with diameters down to submicrons or nanometers can be easily fabricated with electrospinning process. To the mid-1990s, and after the 1990s, this method was investigated intensively.

1.3 Electrospinning Setup

A typical apparatus for electrospinning of polymers consists of a nozzle, a high-voltage power supply, a container for polymer fluid, and an electrode collector of an injection pump with a hypodermic syringe (Fig. 1.2). The solution is pumped through the needle/nozzle, a grounded collector that can be either stationary or rotating. Experiments can also be carried out in a box in order to precisely control environmental conditions such as temperature and relative humidity.

A high DC voltage supplier has two electrodes. One is positive and the other one is negative. The positive end is attached to a polymer solution or a polymer melt, and the negative end is connected to the collecting ground. By adjusting the voltage a required electric field for spinning can be created between the positive and negative electrodes. A polymer fluid—solution or melt—is filled in a capillary tube where a positive electrode wire is inserted. The capillary tube can be a glass capillary, a syringe with a needle, or a nozzle. If a metal needle is used for electrospinning, the positive end is wrapped around the metallic needle. The capillary syringe position can be vertical or horizontal. The polymer fluid holder can be placed horizontally or with various angles. The negative end of the voltage power supplier is connected to a collector opposite the polymer fluid container. Fiber collection screens are generally metallic and covered with an aluminum foil. The shape of the metal collectors is usually flat, but in some cases, for specific fiber production (i.e., aligned fibers) dynamic collectors are utilized instead of stationary ones. Rotating drums, discs, or rotating cylindrical collectors are examples of dynamic screens. Conductive parallel plates are also potential candidates for aligned nanofiber production.

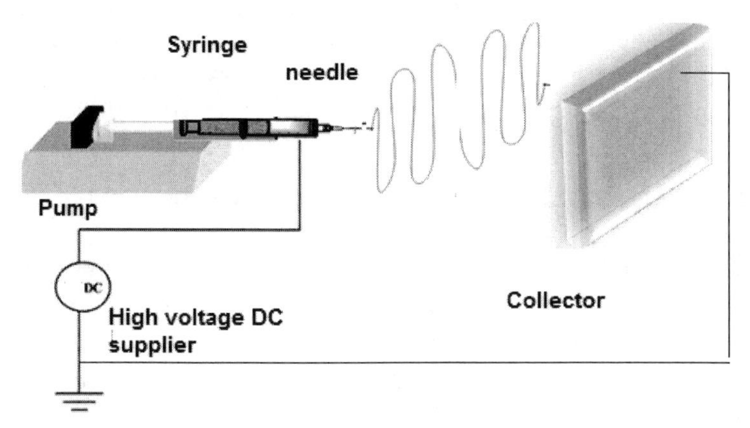

Figure 1.2 A typical setup used to produce nanofibers by electrospinning with an injection pump, a syringe, a high-voltage supply, and a grounded collector.

The electrospinning process has four different stages. In the first stage, an electrically charged liquid polymer jet emerges from the tip of the needle. A whipping process occurs in the second stage. Splaying or multijet formation is accepted as the third stage, and grounding of the thin dried fibers to the collector is the last phase.[17] When an electrostatic force is applied by a high-voltage source, an electric field is formed at the tip of the syringe where the polymer liquid is held by its surface tension. The accumulation of the charges in the tip causes repulsion that opposes the surface tension forces, and the higher the voltage the stronger the mutual repulsion of the charges at the tip.

With the increase of the electric field the pendant polymer drop at the tip of the needle changes its hemispherical shape and takes a conical shape, which is called the Taylor cone. Taylor stated that a conductive liquid can stay in equilibrium with a cone angle of 49.3° under an electric field.[18] Some recent research has shown that the Taylor cone angle is valid for only a specific self-similar solution. A cone angle of 33.5° has been reached both experimentally and theoretically with the initiation of a critical electric field. Surface tension can no longer resist mutual repulsive electrostatic forces, and a charged jet of the polymer solution or melt protrudes from the tip of needle at a point of the Taylor cone. The polymer jet goes through a short stable region and then immediately gains a chaotic motion or instable region starts.[19] In this region solvent evaporation occurs, leaving a thin, dried fiber behind. Fibers are generally collected at the negative polar end as nonwoven mats.

1.4 Uses and Types of Electrospinning

Electrospinning is used for continuous production of electrospun nanofibers webs with high efficiency and discontinuous production of nanofiber webs from a very small amount of liquid (one droplet) for expensive polymers. Application types of electrospinning are the following:[20]

- *Coaxial electrospinning—needle electrospinning*
- *Crosslinking of electrospun nanofibers webs*
- *Bicomponent nanofibers, for drug delivery systems*
- *Needle-less electrospinning*
- *Production of composite materials consisting of electrospun layers, etc.*

Nanofibers have applications in medicine, artificial organ components, tissue engineering, implant material, drug delivery, wound dressing, and medical textile materials. Nanofiber meshes could be used against viruses, and for wound healing at the injury site. Protective materials include sound absorption materials, protective clothings against chemical and biological warfare agents, and sensor applications for detecting chemical agents. Nanofibers have also been used in pigments for cosmetics.[20]

Applications in the textile industry include outerwear garments, sports materials, climbing gear, rainwear, etc. Napkins with nanofibers contain antibodies against various biological hazards, and chemicals that signal by changing color for identifying bacteria.[20]

Filtration system applications include heating, ventilation, and air conditioning (HVAC), high-efficiency particulate arrestance (HEPA), and ultra-low penetration air (ULPA) filters as well as filters for automotive, beverage, pharmacy, and medical applications. Filter media for new air and liquid filtration applications, such as vacuum cleaners, are also applications of filtration systems.[20]

Applications in energy area include Li ion batteries, photovoltaic cells, membrane fuel cells, and dye-sensitized solar cells. Other applications are micropower to operate personal electronic devices via piezoelectric nanofibers woven into clothing, carrier materials for various catalysts, and photocatalytic air/water purification.[19]

1.5 Parameters Affecting Electrospinning

In the electrospinning process, the following three parameter classes have relative effects on the resulting fiber properties:

- *Polymer solution parameters*
- *Polymer processing parameters*
- *Ambient parameters*

Solution conductivity, surface tension, dielectric effect, solution viscosity, which is closely related to the molecular weight of the polymer, solution concentration and polymer chain entanglement, and volatility of the solvent are properties of the spinning solution. Applied voltage (or electrical potential), flow rate of the polymer solution (or feed rate), diameter of the tip, distance between the tip and the collector, and geometry of the collector are the processing parameters.

1.5.1 Polymer Solution Parameters

The solution properties are very effective in the electrospinning process and the produced fiber morphology. During the electrospinning process, the polymer solution will be drawn from the tip of the needle. The electrical property of the solution, viscosity and surface tension will determine the necessary amount is needed for fiber processing solution. The rate of evaporation have an influence on the viscosity of the solution during the stretching. The solubility of the polymer in the solvent will determine the viscosity of the spinning solution at the same time determines the types of polymer which can be mixed together.

1.5.1.1 Surface tension

Surface tension (σ) is defined as force applied to the plane of the surface per unit length. In liquids, a small droplet falling through air takes a spherical shape. The surface property of the liquid that is known as surface tension causes this phenomenon. In the electrospinning process, the polymer solution has to have sufficient charge in order to overcome surface tension in the liquid solution. During electrospinning, beaded fiber formation can be observed within the polymer jet due to the high surface tension values.

Beaded nanofibers were produced from a water/PEO solution. Addition of ethanol to the water/PEO solution reduces the surface tension of the solution, and production of smooth PEO nanofibers can be obtained. High surface tension causes beaded fibers. On the other hand, smooth fibers without bead formation were seen in PVP/ethanol solutions having a lower surface tension. Another way is to add a surfactant to the spinning solution. Surfactant contribution to the spinning solution decreases surface tension. An insoluble surfactant is also used to decrease the surface tension. In addition to solvents and surfactants, temperature is another factor for surface tension. In the pure liquid form, the surface tension of the liquid decrease with increasing temperature, as the equilibrium between the surface tension and the vapor pressure would decrease. At a certain (critical) point, the interface between the liquid and the gas will disappear.

Surface tension (Table 1.2) is a property of the surface of a liquid that allows it to resist an external force, and surface tension is the measurement of the cohesive (excess) energy present at a gas/liquid interface (Fig. 1.3). The molecules of a liquid attract each other. The interactions of a molecule in the bulk of a liquid are balanced by an equally attractive force in all directions. The net effect of this situation is the presence of free energy at the surface. The excess energy is called surface free energy and can be quantified as a measurement of energy/area. It is also possible to describe this situation as having a line tension or surface tension, which is quantified as a force/length measurement. The common units for surface tension are dynes/cm or N/m. Polar liquids, such as water, have strong intermolecular interactions and thus high surface tension. Any factor that decreases the strength of this interaction will lower surface tension. Thus an increase in the temperature of this system will lower surface tension. Any contamination, especially by surfactants, will lower surface tension and lower surface free energy. Some surface tension values of common liquids and solvents are shown in the following Tables 1.2 and 1.3.

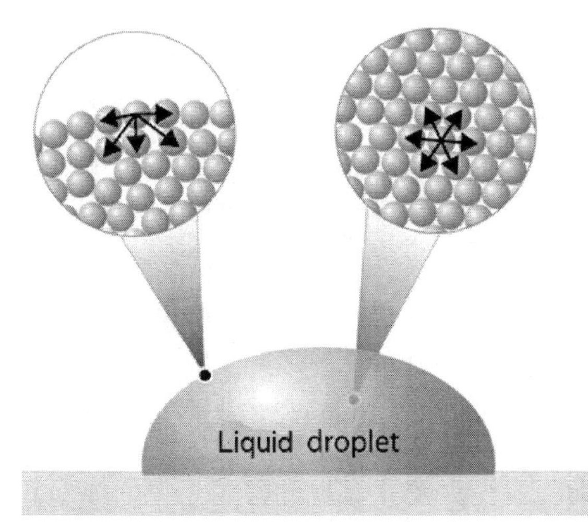

Figure 1.3 Representation of surface tension in a liquid. Molecules at the surface of water experience a net attraction to other molecules in the liquid, which holds the surface of the bulk sample together. In contrast, those in the interior experience uniform attractive forces.

Table 1.2 Surface tension (at 25°C), viscosity, vapor pressure, and boiling points of common liquids

Substance	Surface Tension ($\times 10^{-3}$ J/m^2)	Viscosity (cP) (20°C)	Vapor Pressure, Torr (20°C)	Boiling Point (°C)
Solvents and organics				
Diethyl ether	17	0.24	587	34.6
n-Hexane	18	0.31	160	68.7
Acetone	23.3	0.36	185	56.5
Ethylene glycol	48	16.1	~0.09	198.9
Acetic acid	27.6		15.3	118
n-Heptane	20.14	0.39	35.5	98
n-Hexane	18.4		160	69
o-Xylene	30.03	0.81	6	
Chloroform	27.2	0.57	158.4	61.15
Dichloromethane	28.1	0.44	350	
Methyl ethyl ketone, MEK	24.6	0.43	74	79.64
N,N-dimethylacetamide, DMA	32.43 (30°C)	2.14	1.3 (25°C)	166.1
N,N-dimethylformamide, DMF	37.1	0.92	2.7	153
N-methylpyrrolidone, NMP	40.8	1.67 (25°C)	0.33	202
Dimethylsulfoxide	43.5	2.24	0.6 (25°C)	189
Tetrahydrofuran, THF	26.7	0.46	142	66
Toluene	28.0	0.59	28.5	110.6
Chlorobenzene	33.28	0.80	8.8	131.69
Ethanol	22.3	1.07	59	78.5
Methanol	22.6	0.59	97	64.7
Liquid Elements				
Bromine	41	0.94	218	58.8
Mercury	486	1.53	0.0020	357
Water				
0°C	75.6	1.79	4.6	–
20°C	72.8	1.00	17.5	–
60°C	66.2	0.47	149	–
100°C	58.9	0.28	760	–

Source: Adapted and collected from http://macro.lsu.edu/howto/solvents.htm

The adhesion and uniformity of a film are also influenced by the forces that act between the coating formulation that is in solution form and the core surface of the film-coated surface. The measure of wetting behavior is the contact or wetting angle, which forms between a liquid droplet and the surface of the solid body to which it is applied.

Table 1.3 shows, for example, the surface tension of the solutions of some water-soluble polymers at 25°C (0.1% aqueous solutions of the polymers). Hydroxypropylcellulose (HPC) is a good example of a surface-active polymer. Water solutions greatly reduced surface and interfacial tensions. HPC functions as an assistant in both emulsifying and whipping. HPC combines organic solvent solubility, thermoplasticity and surface activity with the aqueous thickening and stabilizing properties of other water soluble cellulose polymers. The reduction in surface and interfacial tensions of water solutions containing HPC (Table 1.3) as low as 0.01% produces close to the maximum reduction in surface tension.

Table 1.3 Surface tensions of solutions of some water-soluble polymers at 25°C (0.1 wt.% aqueous solution of the polymers; for HPC different percentages of aqueous solutions)

Polymer	Surface Tension (dynes/cm) mN/m
Sodium carboxymethylcellulose (NA-CMC)	71.0
Hydroxyethylcellulose (HEC)	66.8
Hydroxymethylcellulose (HMC)	50–55
Hydroxypropylmethylcellulose (HPMC)	46–51
Hydroxypropylcellulose (HPC) 0 wt.%	74.1
HPC in H_2O 0.01 wt.%	45.0
HPC in H_2O 0.1 wt.%	43.6
HPC in H_2O 0.2 wt.%	43.0

Smaller contact angles give smoother film coatings. The contact angle becomes smaller with decreasing porosity and the film former concentration. Solvents with a high boiling point and a high dielectric constant reduce the contact angle. The higher the critical surface tension of the core, the better the adhesion of the film to the core.

The smaller the contact angle, the better the adhesion of the film to the core.

1.5.1.2 Solution conductivity

For the electrospinning process to be initiated, the solution must gain sufficient charges such that the repulsive forces within the solution are able to overcome the surface tension of the solution. Subsequent stretching or drawing of the electrospinning jet is also dependent on the ability of the solution to carry charges.

Acids, bases, salts, and dissolved carbon dioxide may increase the conductivity of the solvent. The solvent conductivity can be increased remarkably by mixing chemically inert components. Substances, such as mineral salts, mineral acids, carboxylic acids, some complexes of acids with amines, stannous chloride, and some tetraalkylammonium salts, are added to the solvent to increase its conductivity. For organic acid solvents, small amount of water addition will significantly increase their conductivity due to ionization of the solvent molecules. This increase in the conductivity can help production of beadless fibers just because stretching of the solution has increased, and to some degree a fiber diameter decrease can be observed.

1.5.1.3 Dielectric effect

The dielectric constant of a solvent has a significant influence on electrospinning. A solution with a greater dielectric constant reduces bead formation and the diameter of the electrospun fiber.[21] *N,N*-dimethylformamide (DMF) can be added to the electropinning solution to increase its dielectric property to improve the fiber morphology.[22] The bending instability of the electrospinning jet also increases with an increase in the dielectric constant of electrospinning solution. This is shown by the increased deposition area of the fibers, which may also facilitate the reduction of the fiber diameter due to the increased jet path.[23] The dielectric constant and other physical parameters of some common solvents used in electrospinning are given in Tables 1.4 and 1.5.

The relationship between the diameter of the resultant fiber and the dielectric constant of the polymer solution were studied, where resultant fibers from solutions that have a higher dielectric constant have a smaller diameter.

n-butyl acrylate/methyl methacrylate copolymer [P(BAco-MMA)] was synthesized by emulsion polymerization. Nanofibers of P(BA-co-MMA) were produced by electrospinning. The nanofiber mats showed relatively high hydrophobicity with an intrinsic water contact angle up to 120°.[24]

The diameters of P(BA-co-MMA) nanofibers were strongly dependent on the polymer solution dielectric constant, the concentration of the solution, and the flow rate. Homogeneous electrospun P(BA-co-MMA) fibers as small as 390 ± 30 nm were produced. The dielectric properties of the polymer solution strongly affected the diameter and morphology of the electrospun polymer fibers. The bending instability of the electrospinning jet increased with a higher dielectric constant. The charges inside the polymer jet tended to repel each other so as to stretch and reduce the diameter of the polymer fibers by the presence of a high dielectric environment of the solvent. The electrospinning of P(BA-co-MMA) from solvents and mixed solvents very effectively showed how the choice of solvent affects nanofiber characteristics.[24]

Table 1.4 The solvent affects the nanofiber characteristics in the electrospinning of P(BA-co-MMA) from solvents and mixed solvents

Solvents	Ratio of Solvents (v/v)	Dielectric Constant	Diameter of Nanofibers (nm)
DMF	100	36.71	600±20
DMF/THF	75/25	29.40	680±40
DMF/Acetone	50/50	28.71	720±25
DMF/THF	50/50	22.09	800±30
Acetone	100	20.70	870±40
DMF/THF	25/75	14.78	995±50
THF/Acetone	50/50	14.08	1200±40
THF	100	7.47	1600±80, beaded fibers

Source: Reproduced with permission from Ref. 24, Copyright 2013, John Wiley and Sons.

Note: Electrospinning conditions and polymer solution concentrations were the same for all samples. Flow rate: 1 mL/h; distance: 15 cm; applying voltage: 15 kV and 5 wt.% solid.

Table 1.5 Some physical properties of solvents

Solvent	Chemical Formula	Boiling Point	Dielectric Constant	Density (g/mL)	Dipole Moment
Nonpolar solvents					
Pentane	CH_3-CH_2-CH_2-CH_2-CH_3	36°C	1.84	0.626	0.00 D
Hexane	CH_3-CH_2-CH_2-CH_2-CH_2-CH_3	69°C	1.88	0.655	0.00 D
Cyclohexane	C_6H_{12}	81°C	2.02	0.779	0.00 D
Benzene	C_6H_6	80°C	2.30	0.879	0.00 D
Toluene	C_6H_5-CH_3	111°C	2.38	0.867	0.36 D
Chloroform	$CHCl_3$	61°C	4.81	1.498	1.04 D
Polar protic solvents					
Formic acid	H-C(=O)OH	101°C	58.00	1.210	1.41 D
Acetic acid	CH_3-C(=O)OH	118°C	6.2	1.049	1.74 D
Methanol	CH_3-OH	65°C	33.00	0.791	1.70 D
Ethanol	CH_3-CH_2-OH	79°C	24.55	0.789	1.69 D
n-Butanol	CH_3-CH_2-CH_2-CH_2-OH	118°C	18.00	0.810	1.63 D

Solvent	Chemical Formula	Boiling Point	Dielectric Constant	Density (g/mL)	Dipole Moment
Isopropanol (IPA)	CH_3-CH(-OH)-CH_3	82°C	18.00	0.785	1.66 D
n-Propanol	CH_3-CH_2-CH_2-OH	97°C	20.00	0.803	1.68 D
Nitromethane	CH_3-NO_2	100–103°C	35.87	1.137	3.56 D
Water	H-O-H	100°C	80.00	1.000	1.85 D
Polaraprotic solvents					
Dichloromethane (DCM)	CH_2Cl_2	40°C	9.10	1.326	1.60 D
Tetrahydrofuran (THF)	/-CH_2-CH_2-O-CH_2-CH_2-\	66°C	7.50	0.886	1.75 D
Acetonitrile (MeCN)	CH_3-C≡N	82°C	37.50	0.786	3.92 D
Ethyl acetate	CH_3-C(=O)-O-CH_2-CH_3	77°C	6.02	0.894	1.78 D
Acetone	CH_3-C(=O)-CH_3	56°C	21.00	0.786	2.88 D
N,N-dimethylformamide (DMF)	H-C(=O)N$(CH_3)_2$	153°C	38.00	0.944	3.82 D
Dimethyl sulfoxide (DMSO)	CH_3-S(=O)-CH_3	189°C	46.70	1.092	3.96 D
Propylene carbonate	$C_4H_6O_3$	240°C	64.00	1.205	4.98 D

In the electrospinning technique, the ejected charged jet was affected by electrical forces, so it is needed to have high electrical properties, i.e., a good dielectric constant, to enhance the density of charges at the surface of the jet for better stretching and uniform formation of fibers with bead-free morphology.[24]

The copolymer with hydrogen bonding groups displayed an increase in intermolecular associations with a decreasing solvent dielectric constant. [24]

Moreover, strong intermolecular associations between the functional groups were readily observed in the nonpolar solvents with the production of significantly larger electrospun fibers due to an increased effective molecular weight of the polymer chains. The following equation fitted best to the experimental data obtained in the case of P(BA-co-MMA) by nonlinear regression between the average nanofiber diameter denoted as r and the dielectric constant of the solvent or solvent mixture, denoted as ε.[24] Figure 1.4 shows how experimental and calculated values well fitted according to this equation:

$$r = 540 + 2100e^{-0.1\varepsilon}$$

These results led to the conclusion that the average diameter of nanofibers (r) depended on the dielectric constant (ε) of the solvent or solvent mixture through an exponential relationship given above under these study conditions by holding other parameters constant.[24]

1.5.1.4 Solution viscosity

There are several factors affecting solution viscosity. Concentration, molecular weight, polymer chain entanglement, and temperature are accepted as the main factors. The molecular weight of a polymer is directly related to viscosity of the solution. A necessity of electrospinning is that the electrospinning solution should consist of a polymer with sufficient molecular weight and should have sufficient viscosity. As the jet leaves the needle tip during electrospinning, the polymer solution is stretched as it travels toward the collection plate. During the stretching of the polymer solution, it is the entanglement of the molecule chains that prevents the electrically driven jet from breaking up, by maintaining a continuous solution jet. As a result, a monomeric solution does not form fibers when electrospun.[25]

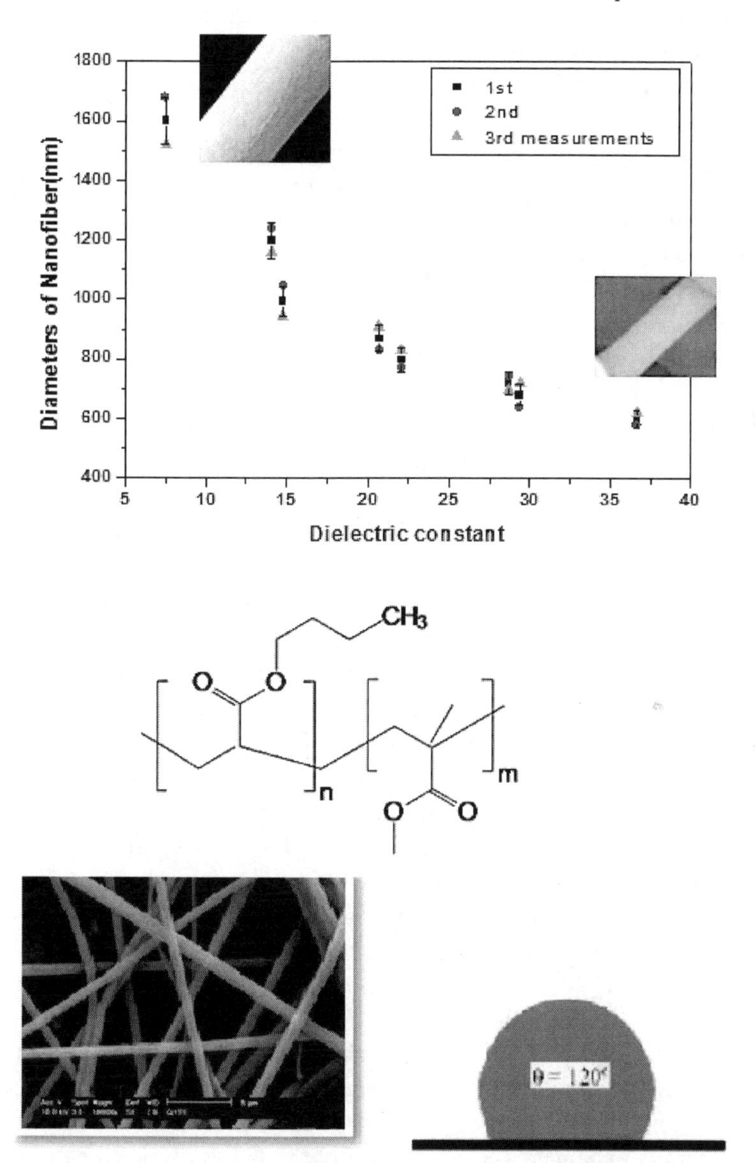

Figure 1.4 Nanofibers of *n*-butyl acrylate/methyl methacrylate copolymer P(BA-co-MMA) were produced by electrospinning. Diameter vs. dielectric constant of the solvent. Reprinted from Ref. 24, Copyright 2013, John Wiley and Sons.

The molecular weight of the polymer represents the length of the polymer chain, which in turn has an effect on the viscosity. Increasing solution concentration shows almost the same effect as using a higher-molecular-weight polymer. Polymer chain entanglement of the polymer solution is improved in either case. At higher concentrations, viscosity of the solution becomes higher and it prevents the jet from having larger bending instabilities. This causes small deposition on the collecting media for fibers and the resultant fiber diameter is thickened. At low viscosities, there will be less amount of chain entanglement in the polymer solution. The forces from surface tension become dominant and bead formation occurs along the string of electrospun fibers. At high viscosities, jets can be stretched fully and beadless fibers can be obtained. High viscosity values also result splitting of jets into smaller fibers, and pumping of the polymer solution becomes difficult, and drying of the solution on the tip of the needle can be observed.

1.5.1.5 Volatility of the solvent

Solvent volatility is an important factor in electrospinning. Since electrospinning requires a quick evaporation rate and phase separation, vapor pressure of the solvent affects the drying time and evaporation rate. Other parameters affecting evaporation rate are boiling point, specific heat, enthalpy and heat of vaporization, rate of heat supply, interaction between solvent molecules, surface tension of the liquid, and air movement above the liquid surface.

Solvent volatility is also an important factor in determining the properties of fibrous structures produced by electrospinning. In the electrospinning process, solvent evaporation occurs while the jet travels from the tip of the syringe to the collector. If all of the solvent evaporates on the way, fibers can be formed and deposited on the collector. However, if some solvent remains on the polymer, instead of dry fibers, wet fibers or thin films can be produced. Solvent volatility might play a role on the formation of pores in the fibers.[26] A decrease in the solvent volatility resulted in a smoother fiber surface. Low-boiling-point solvents are desirable because evaporation of the solvent is enhanced and deposition of the fibers becomes easier. A rapid evaporation rate of the solvent can cause the fibers to form as ribbons with various cross sections.

1.5.2 Polymer Processing Parameters

Other important parameters are the following: the voltage supplied, the feed rate, the solution's temperature, collector's type, the needle's diameter, and the distance between the needle tip and the collector. These parameters can control the fiber morphology but they are less significant compared to the solution parameters.

1.5.2.1 Applied voltage

The applied voltage determines the amount of charges carried by the jet of the polymer. The high voltage will generate the necessary charges on the electrospinning solution and in parallel line with the external electric field which will initiate the electrospinning if the electrostatic force in the electrospinning solution overcomes the surface tension of the solution. In general, a voltage of more than 6 kV can cause the solution drop at the tip of the needle to distort into the shape of a Taylor cone.[27] A higher voltage may be required to stabilize the Taylor cone. The repulsive force in the jet will then stretch the viscoelastic solution. In the case of high voltage application, higher amount of charges will generate faster jet acceleration, and more solution will be going out from the tip of the needle, which may result in a smaller and less stable Taylor cone.[28] When the drawing of the solution to the collection plate is faster than the supply from the source, the Taylor cone may draw back into the needle.[29]

The applied voltage has also effects on the morphology and the resultant fibers. An increase in the applied voltage results in a decrease in the fiber diameter. Generally, a high voltage results in higher bead formation, but increased jet stretching leads to fewer numbers of beads. At lower voltage, due to the weaker electrostatic force, flight time may last longer. Longer flight time lets the jet elongate and stretch stronger and longer, resulting in reduced fiber diameter.

1.5.2.2 Flow rate

The flow rate will determine the amount of solution which will be used for electrospinning. For a certain voltage, there is a corresponding flow rate which maintains the stability of Taylor cone. If the flow rate is increased, there will be a corresponding increase in the fiber diameter or bead size. This is apparent as there is a greater volume of solution that is drawn away from the needle tip. However, there is a limit to the increase in the diameter of the fiber due to the higher flow rate. If the flow rate is at the same rate that the solution is carried away by the electrospinning jet, there should be a corresponding increase in charges when the flow rate is increased. Thus there is a corresponding increase in the stretching of the solution that counters the increased diameter due to increased volume.[30]

1.5.2.3 Distance

The distance between the tip of the needle and the collector has a significant effect on the strength of the electric field and the flying time of the jet throughout the electrospinning path. If the distance between two polar ends is short, solvents may not find the required time to be vaporized entirely before the jet arrives at the collector. The resultant fibers may include some solvents left on them. These residual solvents may cause fibers to stick together and can result in merging of the fibers. A shorter distance between the tip and the collector may lead to an increase in the strength of the electric field. This increase accelerates the velocity of the polymer jets by reducing the travel time of the polymer jet that is electrospun. Reducing the distance does not affect size and shape of the fibers, but inhomogeneously distributed beads can be observed, which might be due to increase in the electric field strength. If the distance between the tip and the collector is longer, the solution jet finds more time for the evaporation of the solvent and the jet can be stretched sufficiently before it forms on the collecting media. Increasing the working distance enhances both the number of beads and the density of the fibers.[31]

1.5.2.4 Effect of the collector

In the electrospinning process, usually a conductive material is used to cover the collecting media. Aluminium foil is one of the most common conductive materials that are used for collection of fibers onto it. By using the conductive material covering, a stable potential difference can be obtained between the tip and the collector. Conductive collectors attract more jets on the surface of the collector, resulting in a higher amount of fiber deposition. For nonconductive collectors, less fiber deposition is seen because charges on the polymer jet flow on the collector due to fast accumulation of the charges. Using a porous collector has also an effect on the resulting fibers. The packing density is usually low in porous collectors. This is mainly due to the rate of evaporation of the residual solvents on the fibers deposited. Fiber morphology can be improved by using a dynamic collector. Rotating cylinders are utilized for the production of aligned fibers. One advantage of using a rotating collector is that solvents have more time for evaporation.

1.5.2.5 Diameter of the needle

In the electrospinning process, the inner diameter of the needle also has some effects. Decreasing the inner diameter can cause an observable decrease in clogging, the number of beads, and the final fiber diameter.

The decrease in the inner diameter creates an increase in the surface tension of the drops on the tip of the needle or pipette, which means that a greater amount of electrostatic force is needed to start the formation of a jet. If no voltage changes occur in the process, jet acceleration decreases and gives more time for stretching and elongating of the jet before it reaches the collector. Too small an inner diameter needle is not preferred due to not being able to form droplets at the tip. It was found that increasing the internal diameter of the capillary increased both the driving voltage necessary for electrospinning and the diameter of fibers deposited on the collecting mesh.

1.6 Theoretical Considerations in Electrospinning

The electrical potential causes the deformation of the fluid drop, and when the applied voltage develops enough force and balances with the fluid surface tension of the polymer solution, the drop is deformed under a cone shape with a semivertical angle of ~30°. Beyond this critical value (Rayleigh limit), the electrostatic forces generated by the charge carriers overcome the surface tension and the deformed droplet undergoes a transition zone just before the fiber jet is initiated to the collector screen. By this way, fluid is submitted to an expressive stretching, but inside tension is still small and a Newtonian flux behavior in this transition zone was suggested. Experimental measurements have shown that typical stretching rates in transition zone are around 100–1000 s^{-1}.

Thus, any change in jet format will indicate a dynamic redistribution of charges on its surface, leading to instability due to bending caused by this redistribution of electric charges. Afterward, a linear segment takes place and the prestretched jet is submitted to rates of 20 s^{-1}. In the linear segment, the flow is basically controlled by effects of electrical field and the longitudinal tension of the viscoelastic fluid.

Due to the high electric fields commonly used in electrospinning, the fluid jet is kept stable under small distances (around 2–4 cm) before reaching the scattering region, where longitudinal instabilities take place. Polymer jets can be ejected at velocities up to 40 $m \cdot s^{-1}$. Figure 1.5 represents the different regions of a polymer jet in the electrospinning process.

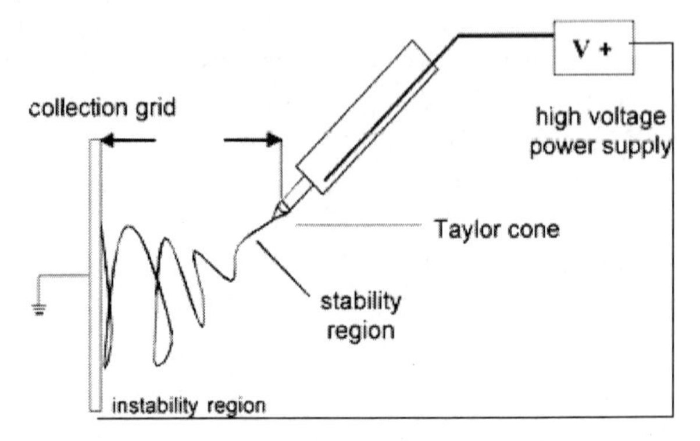

Figure 1.5 Representation of fluid deformation in the electrospinning process.

The electrical instability in the electrospinning process was compared to aerodynamic instabilities where partial differential equations of the aerodynamics of the fluid jet with electric field equations are used to describe the process. Many theoretical models have been used to describe the combination of a viscoelastic fluid driven by an electrical field. On the basis of Taylor studies and with the use of high-velocity cameras, it was proposed that the electrospinning process occurs in two moments, (1) geometry deformation of the fluid droplet by the electric field and (2) formation of a continuous jet from the top of deformed droplet.

The intermediate region between the cone and the scattering region of electrified jets and the fluid jet stability under an electric field have been investigated. The studies searched for evidence of the experimental parameters on the beginning of the electrospinning process. It was suggested that the nanofiber formation is governed by the scattering region, where the surface of charged jets interacts with the electric field, leading to the instability (scattering) region (Fig. 1.5). In other theoretical models, it was considered that the electrospinning jet as a mass-spring system divided into four distinct regions containing beads of electrical charge (e) and mass (m) connected by viscoelastic elements.

For many applications a precise diameter control is required. Fiber dimensions and morphology depend strongly on process parameters such as polymer properties: molar mass, molar mass distribution, glass transition temperature, and solubility; solution properties such as viscosity, viscoelasticity, dielectric constant, concentration, surface tension, electrical conductivity, and vapor pressure; and ambient conditions such as humidity and temperature.

Basically, electrospinning process parameters can be classified into three different types: solution, process, and ambient parameters. Viscosity, conductivity, and surface tension which affect instantly dimensions and morphology of fiber. Viscosity is one of the most important solution parameters.

In the case of homogeneous solutions of a linear polymer, the Huggins equation defines the viscosity of solution:

$$\eta_{sp}(c) = [\eta]\, c + k_H ([\eta]\, c)^2 + \ldots$$

where $\eta_{sp}(c)$ is the specific viscosity, $[\eta]$ is the intrinsic viscosity, c is the polymer concentration, and k_H is the Huggins coefficient. The viscosity and the concentration product, $[\eta]c$, is assigned as the Berry number, B_e.[32]

If a solution to have chain entanglements, B_e becomes bigger than 1. In dilute solutions, polymer chains not overlap each other (Fig. 1.6a), B_e can be unity ($B_e \approx 1$). In the Huggins equation, the intrinsic viscosity is the initial slope of the curve between specific viscosity vs. concentration and is related to the root-mean-squared end-to-end distance, $<R^2>^{1/2}$, of the linear polymer chain by having N monomers related to the Fox–Flory relationship:

$$[\eta] = <R^2>^{3/2}/N$$

According to the Mark–Houwink–Sakurada equation, the intrinsic viscosity $[\eta]$ might be related to the molecular weight (M) of a linear polymer:

$$[\eta] = K.M^a$$

where K and a depend on the polymer, solvent, and temperature. The critical chain overlap concentration, C^*, is the crossover concentration between the dilute and the semidilute concentration regimes. The critical chain overlap concentration is the case if the concentration inside a single macromolecular chain equals the solution concentration and can be defined as

$$C^* \approx N/<R^2>^{3/2} \approx 1/[\eta]$$

From the equation, $C^*[\eta] \approx 1$ in the dilute solution limit, by suggesting the criteria $C^* \approx 1/[\eta]$, as a means of evaluating C^* (since $B_e \approx 1 \Rightarrow C \approx 1/[\eta]$ in the dilute solution limit).

Hence, the calculation of C^* from chain dimensions

$$C^* \approx 3M/4\pi <R^2>^{3/2} N_{av}$$

By considering $C^* \approx 1/[\eta]$, one can estimate C^* (N_{av} is the Avogadro number and M is the molecular weight). In a good solvent, the radius of gyration, R_g, is usually a better estimate of the chain dimensions rather than the root-mean-squared end-to-end distance, $<R_2>^{1/2}$, indicating the hydrodynamic interactions between the polymer chains and the solvent. The radius of gyration can be determined from the hydrodynamic radius, R_h (assuming nondraining conditions), by using Kirkwood–Riseman approach:

$$R_g = R_h/0.875$$

the hydrodynamic radius may be determined by dynamic light scattering measurements.

In the case of dilute solutions of good solvents, the solution viscosity found to be proportional to the concentration, ($\eta \approx C$). This is same with the above equation in the dilute regime, where the power terms of concentration are negligible. However, for $C > C^*$ in good solvents, in the semidilute regime, equation becomes

$$\eta = \eta_s (C/C^*)^{3/(3\nu - 1)}$$

where η_s is the solvent viscosity and the ν is the Flory exponent (0.5 for theta solvents and 0.6 for good solvents). In good solvents, the single parameter scaling model based on the reptation model, where the viscosity scales with molecular weight as $\eta \approx M^{3.0}$, predicts a concentration exponent of 3.75. However, experimental data indicates a stronger dependence of viscosity on molecular weight ($\eta \approx M^{3.4}$) than that predicted by reptation theory. It has been suggested that mechanisms of relaxation other than reptation, such as contour length fluctuations, might account for this observed stronger dependence. If this stronger dependence of η on molecular weight ($\eta \approx M^{3.4}$) is taken into account, then the exponent for the concentration dependence of viscosity has been predicted to be 4.25. A different scaling concept, based on two parameters, was proposed by Colby et al., where an even stronger viscosity dependence on concentration was predicted ($\eta \approx c^{4.5}$) and measured experimentally ($\eta \approx c^{4.8}$).[33]

There are two power law for the semidilute regime, semidilute unentangled and semidilute entangled. A physical representation of the semidilute unentangled regime is represented in Fig. 1.6b, in this regime the concentration is large for chain overlap ($C > C^*$) but not enough for entanglement. By the further concentration increase (semidilute entangled regime, Fig. 1.6c), the crossover of concentration from the semidilute unentangled to the semidilute entangled regime is defined as the critical entanglement concentration, C_e.

As the concentration was increased, droplets and beaded fibers were observed in the semidilute unentangled regime; and beaded as well as uniform fibers were observed in the semidilute entangled regime. Uniform fiber formation was observed at $c/c^* \sim 6$ for all the narrow MWD polymers (M_w of 12, 470–205, 800 g/mol) but for the relatively broad MWD polymers (M_w of 34,070 and 95,800 g/mol), uniform fibers were not formed until higher concentrations, $c/c^* \sim 10$, were utilized.[34]

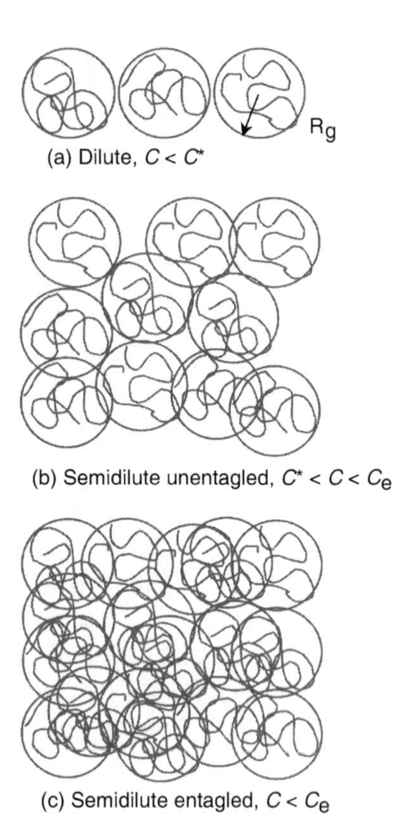

Figure 1.6 Physical representation of the three solution regimes: (a) dilute, (b) semidilute unentangled, and (c) semidilute entangled. Reprinted with permission from Ref. 34, Copyright 2005, Elsevier Ltd. All rights reserved.

1.6.1 Electrospinning of Solutions in the Semidilute Concentration Regime

Above C^* (in the semidilute unentangled regime, $1 < C/C^* < 3$), the morphology was dependent on molecular weight.

The predicted concentration exponents for the viscosity-exponential dependence in a good solvent in the semidilute regime can be expressed as:

$\eta \sim \{C^{1.25}$ semidilute unentangled

$\eta \sim \{C^{4.25-4.5}$ semidilute entangled.

Both solution viscosity and concentration are related by the Berry number, B_e.

Experimental findings show that the diameter of electrospun fibers is dependent on the solution concentration and the polymer chain conformation in solution. Thus, the Berry number is defined by the following equation as indicated in Section 1.6.

$$B_e = [\eta]C$$

where $[\eta]$ is the intrinsic polymer viscosity and C is the polymer solution concentration. As the degree of polymer chain entanglements can be represented by the Berry number, this determines the electrospun fiber diameter. When $B_e < 1$, polymer molecules in solution are sparsely distributed and there is a low probability of an individual molecule to bind with another.

The solution properties play a major role in controlling the diameter and the energy required to electrospun a polymer. The diameter of a spinning fiber is a result of a balance of external forces and the internal resistance to draw. The inherent resistance offered by a spinning polymer solution is in the form of resistance to flow or in the form of its capacity to store elastic energy. It is the elasticity of the system that prevents a polymer from extending infinitely under an applied force. A purely viscous polymer would continue to draw till it become thin and breaks under a constant force. On the other hand, elasticity of the system tends to stabilize the extension of the polymer solution by storing energy by forming a polymer network.[35]

The capacity of a polymer solution to from a network under extension may be judged by the Berry number or the number of entanglements per chain in the solution.[35]

Relationships between the viscosity and concentration in semidilute regimes of linear homopolymers of PMMA in DMF was investigated for electrospun nanofibers (Fig. 1.7). The plot of the zero shear viscosity with the C/C^* distinctly separated into different solution regimes, viz. dilute ($C/C^* < 1$), semidilute unentangled ($1 < C/C^* < 3$) and semidilute entangled ($C/C^* > 3$).

Uniform fibers were formed at high concentrations, i.e., $C/C^* \sim 10$. Dependence of fiber diameter on concentration was determined, i.e., fiber diameter $\sim (C/C^*)^{3.1}$. The fiber diameter was found to vary with the zero shear viscosity of the solutions as,[34]

$$\text{Fiber diameter} \sim \eta^{0.71}.$$

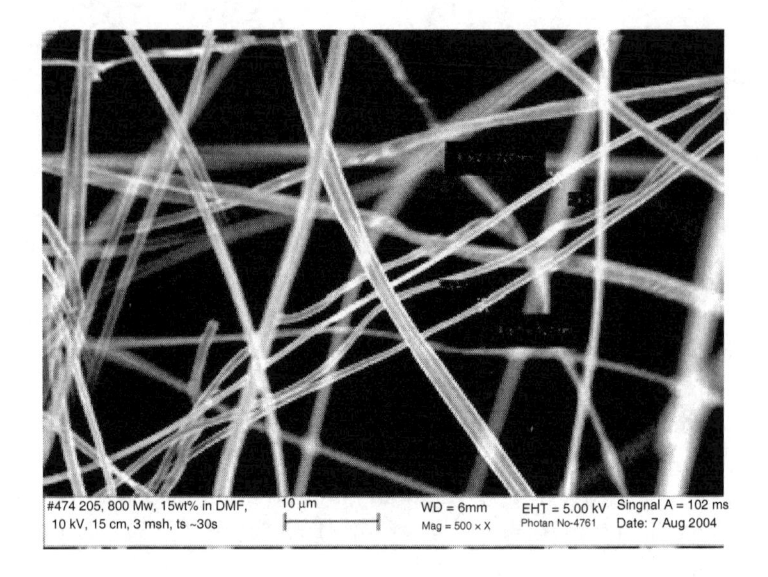

(a)

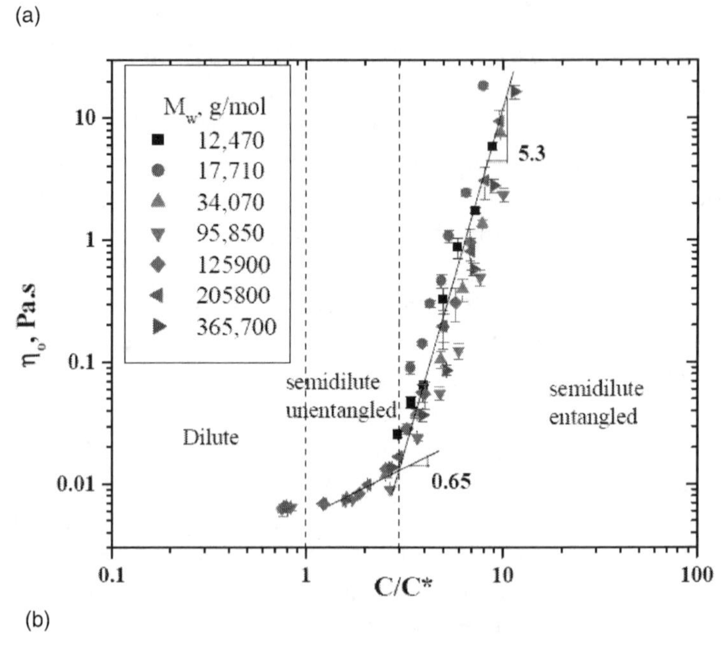

(b)

Figure 1.7 (a) SEM of $C/C_e > 6$, semidilute entangled (205, 800 M_w; $C/C^* \approx 6.8$). (b) Plot of zero shear rate viscosity, η_0, with C/C^* for the different M_w grades of poly(methyl methacrylate) (PMMA). Reprinted with permission from Ref. 34. Copyright © 2005, Elsevier Ltd.

A series of poly(acrylonitrile-co-methylacrylate) copolymers with varying molecular weights were dissolved at different concentrations in DMF and electrospun at a minimum electrospinning voltage (MEV) to correlate the electrical energy required to perform the mechanical work during the spinning of the fibers.

> The diameters of the resultant fibers were correlated with the Berry number and the average number of entanglements per chain of the spinning solution. It was observed that the number of entanglements per chain, which represents the capacity of the polymer system to store elastic energy, could correlate well with the ultimate diameter of the fibers. The diameters of the nanofibers were found to increase linearly with an increase in the number of entanglements per chain with two distinct regions having transition of the slope at the entanglement number value of 3.5.*

In the case of evaporation of the solvent, the extruded polymer solution properties inside the spinning zone would vary in relation to the original properties of the solutions taken and would, therefore, correlate with the original concentration and molecular weight of the original polymer solution. Figure 1.8 shows the relationship of nanofiber diameter against the Berry number, which is the product of $[\eta]C$.[35]

It was observed that the electrospinning of polymer solution below the Berry number of 10 yielded droplets or fibers with droplets and the nanofiber diameter increased with an increase in the Berry number from 65 nm to above submicron and micron size when the Berry number increased above ~10. The data can be fitted using a set of straight lines with different slopes in two different regions (Fig. 1.8). The correlation coefficient for the first trendline improved from 0.891 for the Berry number plot to 0.974 for the number of entanglements plot.[35]

For PLA electrospun fibers beadless at $1 \leq B_e \leq 2.7$, entanglement probability increases and favorable conditions for fiber production takes place. This Berry number range is convenient for nanofiber production of PLA.

At $B_e \geq 2.7$, polymer chain entanglements probability increases and the average fiber diameter goes above the micrometer range.

Surface tension is directly related to the Taylor cone formation and this is related to the electric field strength applied over the fluid droplet able to deform its shape (Fig. 1.9). This tension value is

---

*Excerpt reprinted with permission from Ref. 35, Copyright 2013, The Korean Fiber Society and Springer Science+Business Media Dordrecht.

called critical tension, and polymer solutions with different solvents will have different critical tension values.

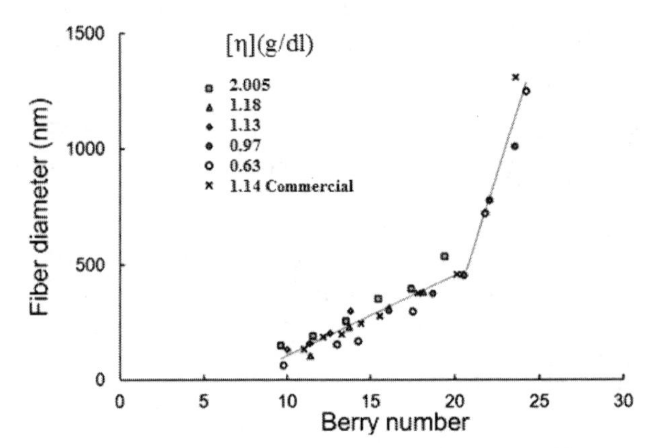

Figure 1.8 Plot of nanofiber diameter against the Berry number. Reprinted with permission from Ref. 35. Copyright © 2013, The Korean Fiber Society+Business Media Dordrecht.

The decrease in surface tension values will favor a beadless morphology. The electrical conductivity of solution also plays an important role in fiber morphology. Higher solution electrical conductivity is associated with a greater number of charges in solution, which favors the electrospinning process. Generally, both electrical conductivity of solvents and polymers are small, and in some cases, inorganic salts are added to the solution, which favor the spinning process. By this method uniform diameter values with beadless morphology were obtained with narrow fiber diameter distribution for poly(ε-caprolactone) (PCL) via electrospinning.

1.7 Effect of Polymer Concentration and Filler

It was found that the resulting fiber diameter was highly dependent on the stretching and acceleration of the fiber jet prior to solidification.[36] A model was developed to predict fiber diameter related to the surface tension and electrostatic charge repulsion. In this model, solution conductivity and viscosity are also taken into account.[37] Among the parameters concerned, concentration/solution viscosity is one of the most important factors affecting fiber diameter. Figure 1.10 displays the effect of polymer concentration on the fiber diameter of PCL fibers prepared from DMF and methylene

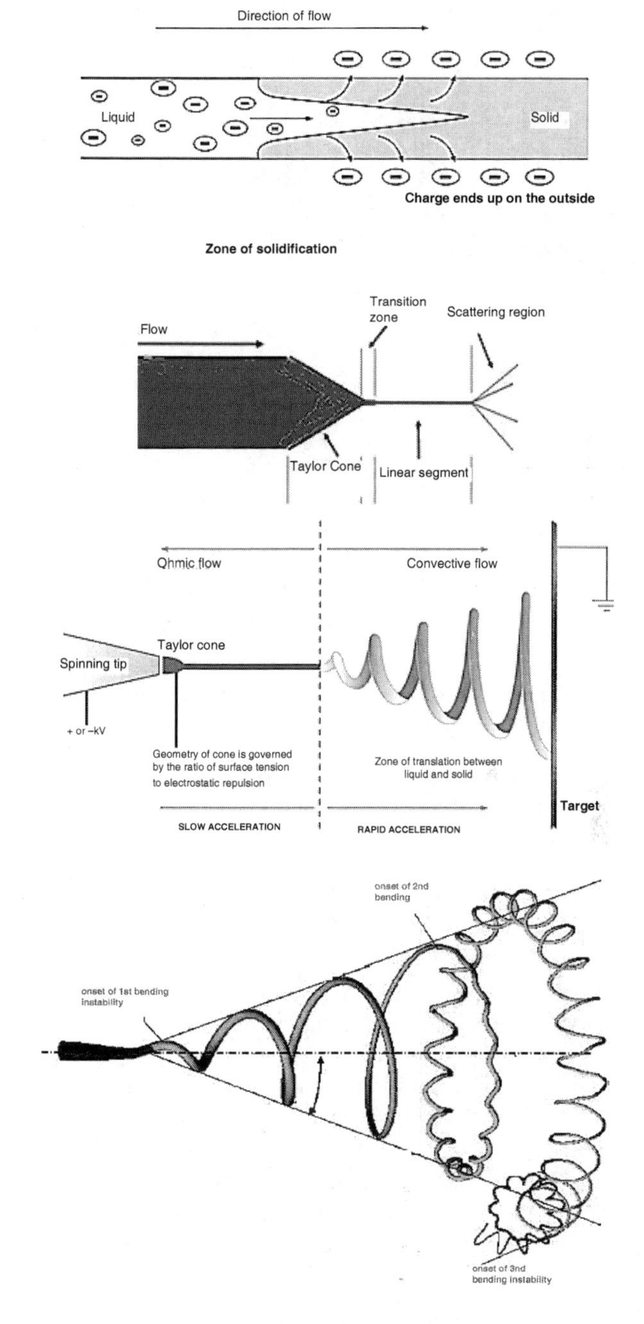

Figure 1.9 Zone of solidification and Taylor cone formation.

chloride, where the fiber diameter increases significantly from 10–20 wt.% PCL to 40 wt.% PCL.

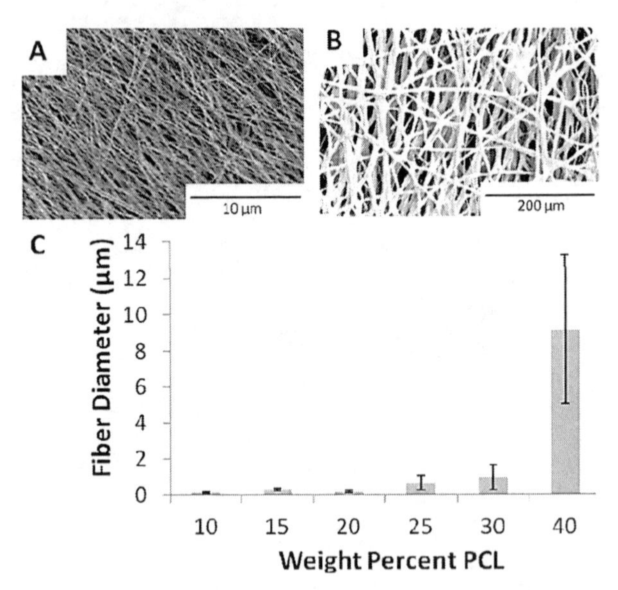

Figure 1.10 Scanning electron micrographs of electrospun poly(ε-caprolactone) (PCL) fibers formed from solutions of varying PCL concentrations in methylene chloride and *N,N*-dimethylformamide (DMF) (50/50 w/w) solvent. (A) 20 wt.% PCL, (B) 40 wt.% PCL, and (C) chart of PCL concentration (wt.%) and fiber diameter (as mean ± standard deviation). Reprinted from Ref. 38, © 2013 by MDPI.

A solvent system that involves an acid–base reaction to produce weak salt complexes is used, which serves to increase the conductivity of the polymer solution.

In electrospinning, process parameters are typically considered as the applied electric field, working distance, flow rate, and, in some cases, collector (rotor) velocity.

There is a range of applied voltage values where a stable jet is obtained for PEO solutions. For example, in solutions at 6 wt.%, a stable jet is formed between 5 and 15 KV, with a working distance of about 12.5 cm. It is also possible to obtain poly(acrylonitrile-co-vinylacetate [P(AN-co-VAc)] nanofibers embedded with magnetic Fe_2O_3 nanoparticles by the use of P(AN-co-VAc)/Fe_2O_3 core–shell nanocapsules applying 15 kV with the feed rate of 0.6 mL/h and the distance of 15 cm. (Fig. 1.11).[39]

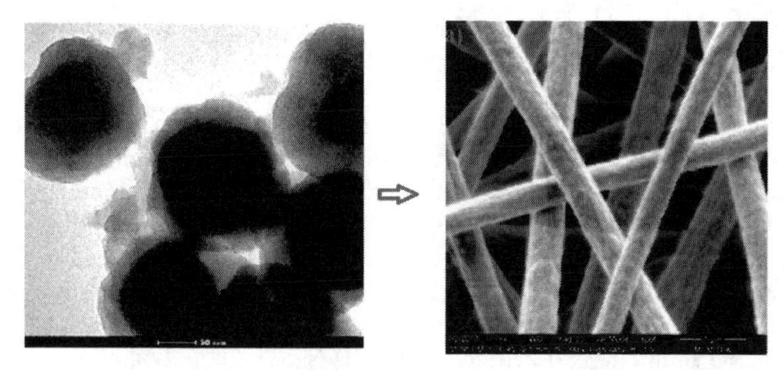

Figure 1.11 Magnetic nanoparticles inside polymeric nanofibers. Transmission electron microscopy (TEM) image of P(AN-co-VAc)/Fe_2O_3 nanoparticles and nanofiber embedded with magnetic nanoparticles. Reprinted with permission from Ref. 39, Copyright 2015, Taylor & Francis.

The diameter of the jet decreases as it moves away from the needle tip until it reaches a minimum value dependent on the Taylor cone initiated. These values were also supported by using PVA, and applied voltage values are directly related to bead formation, and the monitoring of electric current during the process is able to indicate the electric field values where bead density significantly increases.

> By considering axisymmetric instabilities of highly conducting viscoeleastic solutions of PEO, in the theoretical studies, a linear stability analysis combined with a model for stable electrospun jet was used to calculate the expected bead growth rate and the bead wave number for given electrospinning conditions.
>
> The analysis reveals that the unstable axisymmetric mode for electrically driven, highly conducting jets is not a capillary mode but is mainly driven by electrical forces due to the interaction of charges on the jet. Both experiments and stability analysis elucidated that the axisymmetric instability with a high growth rate can be seen in practice when the electrical force is effectively coupled with viscoelastic forces.*

Unlike the applied electric field, the working distance, that is, distance between the needle tip and the collector, seems less important in the formation and morphology of fibers. But a value of the minimum working distance is needed to ensure complete solvent

*Excerpt reprinted with permission from Ref. 40, Copyright 2009, AIP Publishing LLC.

evaporation, and a maximum value for the electric field is effective in forming the Taylor cone and consequently nanofibers.

Fiber diameter decreases when the working distance increases from several centimeters to ~15 cm. Further increase in the distance, from 15 to 20 cm, does not have much effect on fiber diameter. Also, in small working distance conditions, the solvent is not completely evaporated when fibers reach the collector and porous morphologies are obtained. The environmental parameters, temperature, humidity, and air composition can affect the formation and morphology of nanofibers.

Acrylonitrile-co-itaconic acid (AN–IA) copolymers were synthesized in an aqueous medium, changing the monomer feed ratios.[41.] Copolymerizations were achieved with high conversion owing to water-phase precipitation polymerization.

The IA-1% sample, which has the lowest IA content, exhibited the highest average nanofiber diameter, 878 ± 18 nm, and the diameter of fibers was reduced to 386 ± 18 nm for 9% IA content (Figs. 1.12 and 1.13). The average nanofiber diameter decreases corresponding to the IA content in copolymers. A decrement in the standard deviation of nanofibers shows that uniformity of nanofibers is improved by increasing IA.

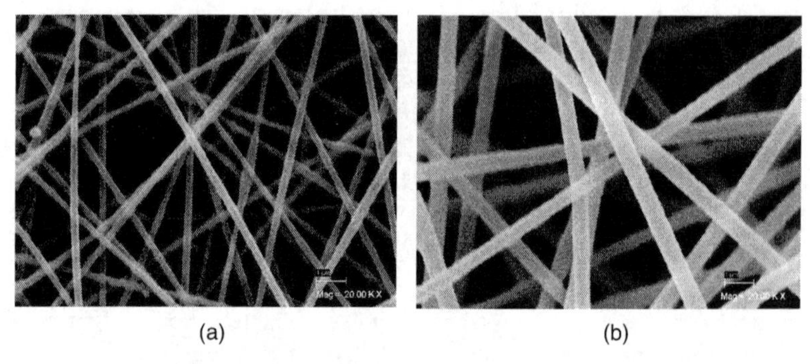

Figure 1.12 SEM images of electrospun nanofibers of P(AN-co-IA) with 20,000 magnitute. (a) IA-9% and (b) IA-1%. Reprinted from Ref. 41, Copyright 2013, Journal of Advances in Chemistry.

Increasing IA concentration in copolymerization results in lower viscosities (Fig. 1.14). IA content of copolymers was determined using Fourier transform infrared spectroscopy–attenuated total reflectance (FTIR-ATR) and nuclear magnetic resonance (NMR) spectroscopies. The absoption bands at 1628 and 1730 cm^{-1} due to C=O stretching vibrations in carboxylic acid showed an increase comparing to the characteristic peaks of the C≡N stretching vibration of AN in copolymers that had higher IA content. ^{1}H-NMR results also supports AN–IA compositions in copolymers.

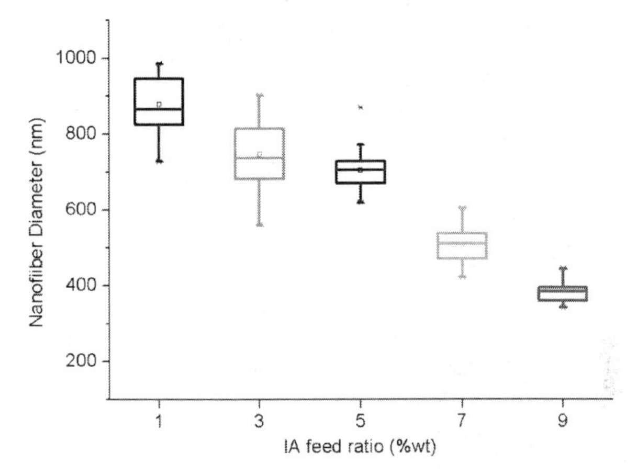

Figure 1.13 Nanofiber diameter range for AN–IA copolymer. Reprinted from Ref. 41, Copyright 2013, Journal of Advances in Chemistry.

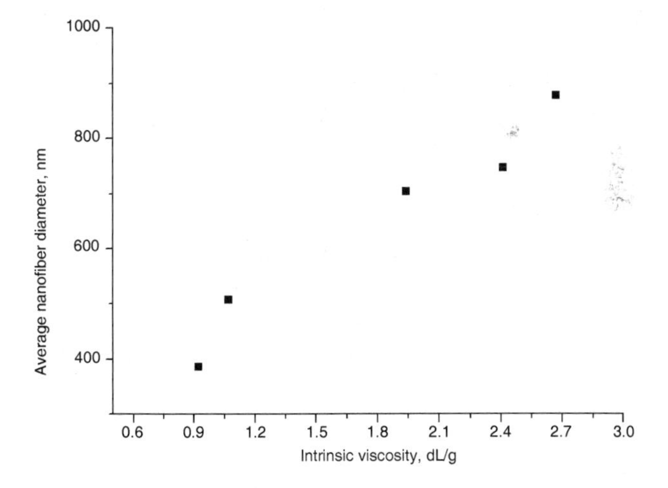

Figure 1.14 Relation between nanofiber diameter and intrinsic viscosity. Reprinted from Ref. 41, Copyright 2013, Journal of Advances in Chemistry.

IA content confirmed by spectroscopic analysis affects thermal properties, which is important for carbon nanofiber production. On the basis of differential scanning calorimetry (DSC) studies, IA provides a catalytic effect on the stabilization process by decreasing the initiation cyclization reaction temperature from 202°C to 195°C. Oxidative reactions are also decreased because of increasing acid content. Thermal gravimetric analysis (TGA) shows two reactions based on weight loss. Thermal decomposition reactions of releasing HCN, NH_3, CO_2, and H_2O gases starts at the second step. The copolymers having high IA content shows decomposition at lower temperatures. An IA comonomer having a crystalline structure could not show an improving effect on segmental mobility of macromolecule chains, and the T_g values increases.

Elecrospinning from the AN–IA copolymer solutions in DMF was performed and morphology of nanofibers was monitored using scanning electron microscopy (SEM).

The nanofiber mats produced were treated at high temperature under an air atmosphere to investigate physical and chemical changes occurring on the copolymers. The color of nanofiber mats turned to different shades of brown-red from white visibly after the heat treatment. Stabilized nanofibers were characterized using FTIR-ATR spectrometry and a new structure was monitored as a result of cyclization reactions. The nanofibers produced from the copolymer having the smallest IA content showed the highest volume loss, 48.7% which is obtained from thermal analysis. According to DSC and TGA results the copolymers synthesized with an IA feed higher than 1% (wt.) have exhibited good stabilization performance (Table 1.6).

Table 1.6 The diameters of stabilized nanofibers

Sample	Avg. Nanofiber Diameter (nm)	Volume Loss (%)
C(IA-1%)	629±13	48.7
C(IA-3%)	619±9	31.3
C(IA-5%)	738±11	21.3
C(IA-7%)	434±7	27.0
C(IA-9%)	341±6	21.9

Source: Reproduced with permission from Ref. 41.

Consequently, electrospun nanofibers of AN–IA copolymers were proposed as a carbon nanofiber precursor due to suitability for electrospinning and the stabilization process.

Electrospun fibers of PVA, PLA, poly(vinyl chloride) (PVC), chitosan, and PS had their morphology strongly dependent on the relative humidity surrounding the spinning process (Fig. 1.15). Depending on the relative humidity used, fibers with different sizes and porosities were obtained.

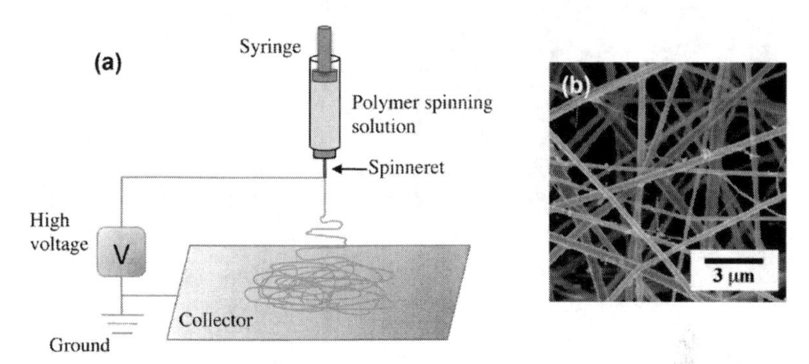

Figure 1.15 Chitosan nanofibers produced by electrospinning. (a) Illustration of a typical vertical electrospinning system for producing polymer nanofibers. (b) SEM image of pure chitosan nanofibers fabricated by electrospinning. Reprinted with permission from Ref. 42, Copyright © 2004 WILEY-VCH Verlag GmbH & Co. KGaA, Weinheim, and from Ref. 43, Copyright © 2014, Royal Society of Chemistry.

FESEM images of the chitosan (CS)/PVP electrospun composite nanofibers[44] mat surface are shown in Fig. 1.16. Obtained fibers have shown uniform structures without any "bead on a string" morphology. They have smooth surfaces, with no particles separating out from the nanofiber matrix, and the fibers have an average diameter of 77 ± 10 nm. Transmission electron microscopy (TEM) images of the nanofibers (Fig. 1.16) suggested that they had a homogeneous inner structure reflected by the uniform gray color of both the upper and lower nanofibers, where CS molecules are evenly distributed throughout the PVP matrix.[44]

The effect of humidity and temperature on the nanofibers of cellulose acetate and PVP was also investigated. It was found that for PVP, increased humidity resulted in a decrease in the average fiber diameter, while for cellulose acetate fiber diameter increased due to the chemical nature of the polymer. However, the dependence of the diameter on temperature was not linear for both polymers, since for lower temperatures, 10°C and 20°C, initially there was an increase in diameter. An increase in temperature to 30°C resulted in a decrease in diameter.

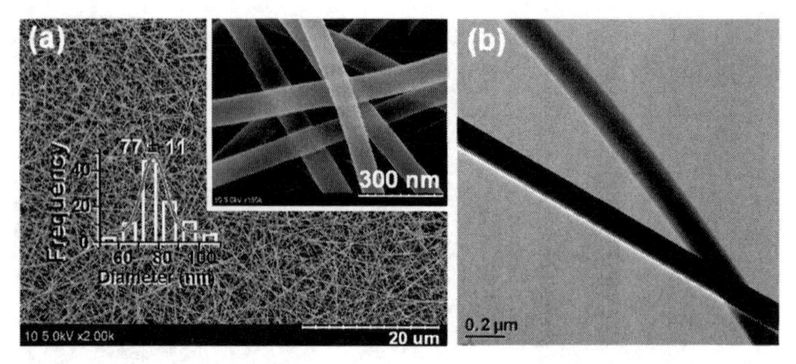

Figure 1.16 Morphologies of electrospun nanofibers. (a) FESEM image and (b) TEM image. Reprinted from Ref. 44. Copyright 2012 Yu DG, et al.

The dielectric constant, viscosity, and surface tension of the solvents affect the electrospinnability, morphological appearance, and fiber size. It was also observed that small and uniform PVP fibers can be obtained using solvents with high dielectric constants, low surface tension, and low viscosity. Furthermore, diameters of PVP fibers decreased with the dielectric constant, dipole moment, and density of the solvents.

A coaxial setup uses a multiple solution feed system that allows for the injection of one solution into another at the tip of the spinneret. The sheath fluid act as a carrier drawing in the inner fluid at the Taylor cone of the electrospinning jet. If the solutions are immiscible then a core–shell structure is usually observed. Miscible solutions, however, can result in porosity or a fiber with distinct phases due to phase separation during solidification of the fiber.

The electrically charged part and coaxial jet shown in Fig. 1.17 allow to obtain core–shell nanofibers (nanochannels and nanocapsules) by coaxial electrospinning.[45]

1.8 Conductive Nanofibrils by Coating Cellulose

Nanowhiskers with different thicknesses of PANI were obtained. One of the advantages of using these coated whiskers instead of pristine conjugated polymers is the inherent strong nature of cellulose allied with the conductive nature of PANI.

A technique called solution blow spinning has been developed as an alternative method for making nonwoven webs of micro- and nanofibers down to the nanometer scale with the advantage of having a higher fiber production. This solution blow spinning method is based on the use of a syringe pump to deliver a polymer solution connected to an apparatus consisting of concentric nozzles. The polymer solution is pumped through the inner nozzle, while a constant high-velocity gas flow is sustained through the outer nozzle (Fig. 1.18).

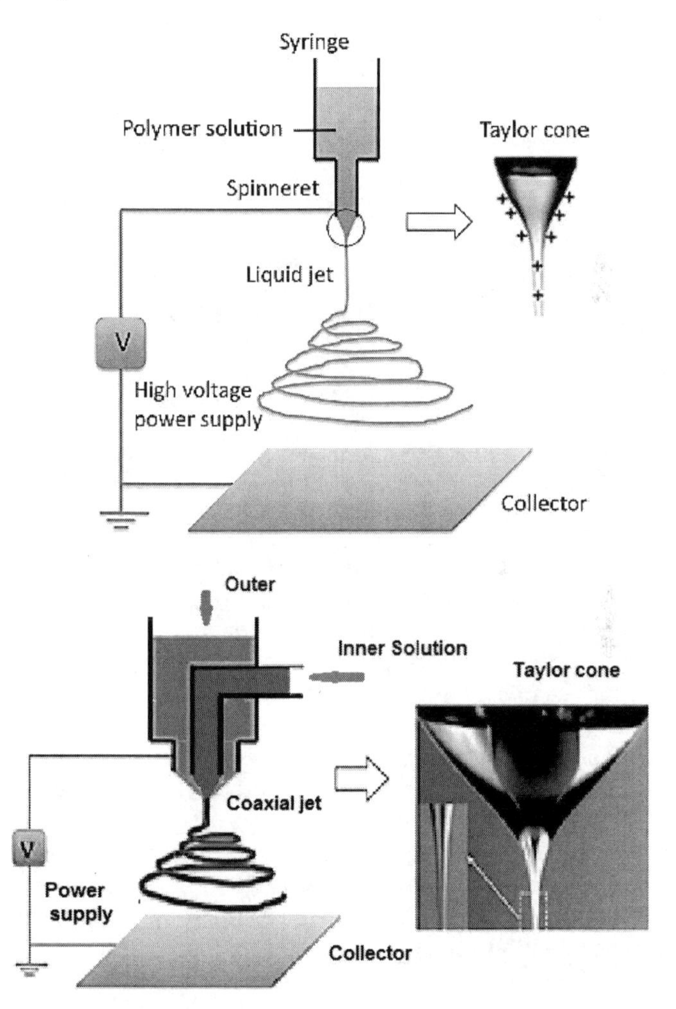

Figure 1.17 Common setup and electrically charged part for coaxial jet for electrospinning and core–shell nanofibers: nanochannel and capsule by coaxial electrospinning. Reprinted from Ref. 45, Copyright 2010, Fengyu Li, Yong Zhao, and Yanlin Song.

The aerodynamic forces are able to stretch the solution exiting the inner nozzle to produce long filaments with a diameter down to the nanoscale.

With this technique the possibility of the production of different morphologies was suggested, that is, smooth and porous fibers as well as beaded fibers. Moreover, polymers such as PLA, PMMA, PEO, PS, and PVC have been used to produce micro- and nanosized fibers with diameters as low as 40 nm.

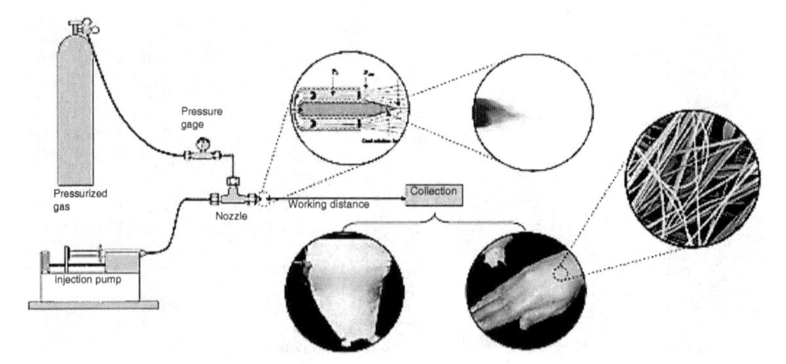

Figure 1.18 Schematic illustration of the solution blow spinning technique. Reprinted with permission from Ref. 46. Copyright © 2009 Wiley Periodicals, Inc.

Another technique to obtain electroactive fibers is through the application of a static electric field on a polymer solution or melt. This technique can be used to produce polymer fibers down to the nanometer scale. In contrast to the common fiber processing techniques such as melt spinning, dry spinning, wet spinning, and extrusion, electrospinning is able to produce ultrathin fibers at low cost and with an elevated surface area, by giving the opportunity for the production of sensors, actuators, and other electroactive devices. MacDiarmid et al. firstly verified the possibility to produce electrospun fibers of conducting polymers once no chain degradation was observed after electrical field application. Since then, many conducting polymers such as PANI, polypyrrole (PPy), and polythiophene have been used to produce nanosized electrospun fibers, including in the form of blend in polymer matrice or as core shell structures of conjugated polymer with nonconjugated polymer matrice.[47–50]

Although there are enough literature on electrospinning of polymer mats, but limited reports can be seen on electrospinning of electroactive polymers and its related applications.

Chapter 2

Applications of Nanofibers

2.1 Applications of Nanofibers

In the constructions made by nanofibers, the high volume-to-weight ratio, soft handling, and high strength to form a barrier to microorganisms and small particles are the main reasons for using them in many applications. These advantages of nanofibers make them sparingly appealing for a broad array of potential applications in many industry segments. Nanofiber applications are shown in Fig. 2.1 and in Table 2.1.

2.1.1 Filtration Applications

Filters have been extensively used in households and industry for removing particles from air or liquid. In the case of enviromental protection filters are used to remove pollutants from air or water. In military, filters are used in uniform garments and isolating bags to decontaminate aerosol dusts, bacteria, while maintaining permeability to moisture vapor for comfort. A respirator is another example requiring an efficient filtration function. The same properties are also needed for some fabrics used in the medical field.[51]

Nanofibers of Conjugated Polymers
A. Sezai Sarac

ISBN 978-981-4613-51-4 (Hardcover), 978-981-4613-52-1 (eBook)
www.panstanford.com

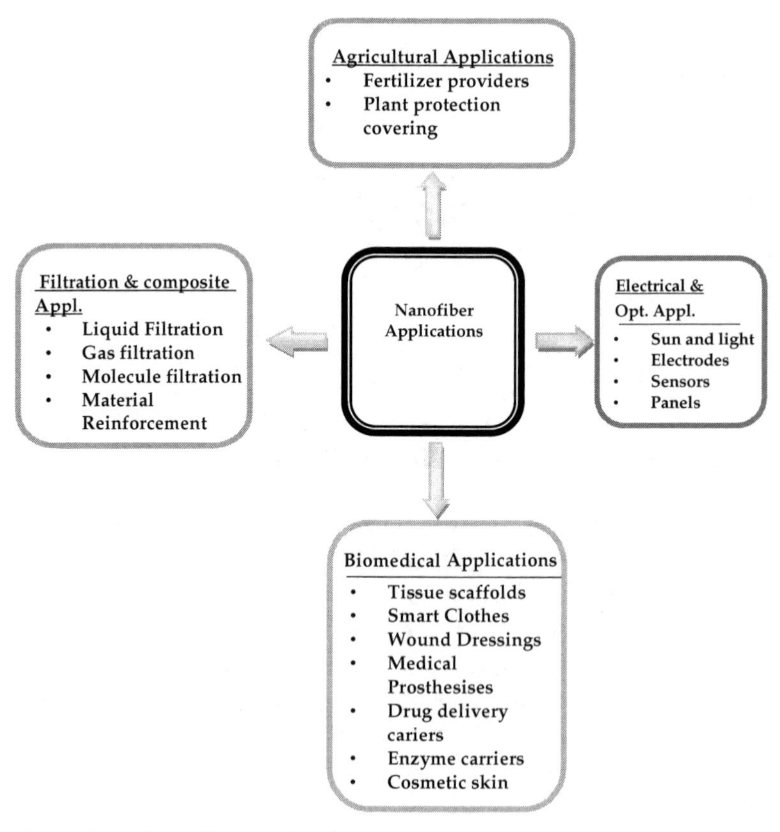

Figure 2.1 Nanofiber applications.

For a fiber-based filter, removal of particles is determined by different mechanisms. Large particles are blocked on the filter surface due to the sieve effect. Particles that are smaller than the surface pores will penetrate into the filter, which could still be collected by the fibers, via either interception or impaction or static electrical attraction.[51]

2.1.2 Nanocomposites

The use of reinforcing fillers and fibers in polymers to improve their mechanical properties is commonly encountered in polymer technology. Conventional fibers such as carbon fibers, glass fibers,

Table 2.1 Electrospun polymers from solution

	Polymer	M_w	Solvent	%	Application
1	Nylon-6,6,-polyamide-(PA-6,6)		FA	10 wt.%	Protective clothing
2	Polyurethane, PU		DMF	10 wt.%	Protective clothing
			DMF	10 wt.%	Electret filter
3	Polybenzimidazole, PBI		DMAc	10 wt.%	Protective clothing, reinforced composites
4	Polycarbonate, PC		DMF:THF (1:1) (1:1)	10–15 wt.%	Protective clothing
			DCM	15 wt.%	Sensor, filter
		$M_w = 60{,}000$	$CHCl_3$, THF		
			DMF:THF (1:1)	20 wt.%	Electret filter
6	Polyacrylonitrile, PAN		DMF	6–15 wt.%	Textile coating
			DMF		Carbon nanofiber
7	Poly(vinyl alcohol), PVA	$M_n = 65{,}000$	H_2O	8–16 wt.% 4–10 wt.%	
		$M_n = 150{,}000$		1–10 wt.%	
8	Poly(lactic acid), PLA	poly(D,L-lactic acid) $M_w = 100{,}000$	DCM and DMF		Membrane for prevention of surgery induced-adhesion

(*Continued*)

Table 2.1 (*Continued*)

	Polymer	M_w	Solvent	%	Application
		poly(L-lactic acid) M_n = 150,000	DCM	5 wt.%	Sensor, filter
		M_w = 205 kDa		14 wt.%	Drug delivery system
9	Poly(ethylene-co-vinylacetate) PEVA	M_w = 60.4 kDa		14 wt.%	Drug delivery system
10	PEVA/PLA	PEVA/PLA=50/50		14 wt.%	Drug delivery system
11	Poly(methyl methacrylate), (PMMA)/tetrahydro-perfluoro-octylacrylate (TAN)	0%–10% TAN	DMF: toluene (1:9)		
12	Poly(ethylene oxide), PEO	M_w = 400,000 and 2,000,000	H_2O [IP: H_2O (6:1) M :300K]	7–10 wt.%	
		M_w = 9.10^5 –1.10^6	H_2O and ETOH or NaCl, H_2O: ETOH(3:2)	1–4.3 wt.%	Micro-electronic wiring, interconnects
		M_w = 300,000	H_2O, $CHCl_3$, acetone,IP		
		M_n = 58,000–100,000		1–10 wt.%	
			IP+ H_2O,	10 wt.%	Electret filter
		M = 100 K to 2 M	$CHCl_3$	0.5–30 wt.%	

	Polymer	M_w	Solvent	%	Application
13	Collagen-PEO	Purified collagen, M_w = 900 kDa	HCl	1–2 wt.%	Wound healing, tissue engineering, hemostatic agents
		PEO: M_n = 900,000	HCl (pH =2.0)	1 wt.%	Wound healing, tissue engineering
14	Polyaniline (PANI)/PEO blend		$CHCl_3$		Conductive fiber
			CSA	2 wt.%	Conductive fiber
		PANI: M_w = 120 kDa, PEO: M_w = 900 kDa, PANI/CSA /PEO: 11–50 wt.%	$CHCl_3$	2–4 wt.%	Conductive fiber
15	PANI/polystyrene (PS)		$CHCl_3$		Conductive fiber
			CSA	2 wt.%	Conductive fiber
16	Silk-like polymer-with fibronectin functionality		FA	0.8–16.2 wt.%	Implantable device
17	Poly(vinyl carbazole)	M_w = 1,100,000	DCM	7.5 wt.%	Sensor, filter
18	Poly(ethyleneterephthalate), PET	M_w = 10,000–20,000	DCM and trifluoracetic	4 wt.%	
			DCM:trifluoroacetic acid (1:1)	12–18 wt.%	

(*Continued*)

Table 2.1 (*Continued*)

	Polymer	M_w	Solvent	%	Application
19	Poly(acrylic acid)–poly(pyrene methanol), PAA-PM	$M_w = 50,000$	DMF		Optical sensor
20	Polystyrene, PS	$M_w = 190,000–280,000$	THF, DMF, CS_2, toluene,	15–35 wt.%	Flat ribbons, catalyst, filter
		$M = 200$ kDa	MEK	8%	Enzymatic biotrans-formation
			$CHCl_3$, DMF	2.5%–10.7%	
		$M_w = 280,000$/ $M_w = 28,000$: 90/1 to 90/10 ratios	THF	15 wt.%	Catalyst, filter
21	Poly(methylmethacrylate), PMMA	$M_w = 540,000$	THF, ACTN, $CHCl_3$		
22	PA		DMAc		Glass fiber filter media
23	Silk/PEO blend	M_w (PEO) = 900,000	Silk /H_2O	4.8–8.8 wt.%	Biomaterial scaffolds
24	Poly(N-vinylpyrrolidone), PVP	$M_w = 20,000$, 100,000	THF	20, 60% (w/v)	Antimicrobial agent

	Polymer	M_w	Solvent	%	Application
25	Poly(vinyl chloride), PVC		THF/DMF=100/0, to 0/100 (vol.%)	10–15 wt.%	
26	Cellulose acetate, CA		ACTN, HOAc, DMAc	12.5%–20%	Membrane
27	Mixture of polyacrylic acid–poly(pyrene methanol) and polyurethane		DMF	26 wt.%	Optical sensor
28	Polyvinylalcohol (PVA)/Silica,	PVA: M_n = 86,000, silica content (wt.%): 0, 22–59	H_2O		
30	PLGA	PLGA(PLA/PGA) = (85/15)	THF:DMF (1:1)	1 g/20 mL	
31	Collagen		HFIP		
32	Poly(ε-caprolactone), PCL		$CHCl_3$: MEOH (3:1) toluene:MEOH (1:1), &CH_2Cl_2:MEOH (3:1)		
33	Poly(2-hydroxyethyl methacrylate), HEMA	M = 200,000	ETOH:FA(1:1), ETOH	12, 20 wt.%/8, 16, 20 wt.%	
34	Poly(vinylidene fluoride), PVDF	M = 107,000	DMF:DMAc (1/1)	20 wt.%	

(Continued)

Table 2.1 (*Continued*)

	Polymer	M_w	Solvent	%	Application
35	Polyether imide, PEI		HFIP	10 wt.%	
36	Poly(ethylene glycol), PEG	M = 10 kDa	$CHCl_3$	0.5–30 wt.%	
37	Nylon-4,6, PA-4,6		FA	10 wt.%	
38	Poly(ferrocenyl-dimethylsilane), PFDMS	M_w = 87,000	THF:DMF (9:1)	30 wt.%	
39	Nylon-6 (PA-6)/ Montmorillonite (Mt)	M_t content = 7.5 wt.%	HFIP, HFIP/DMF:95/5 (wt.%)	10 wt.%	
40	Poly(ethylene-co-vinyl alcohol)	Vinyl alcohol repeat unit: 56–71 mole%	IP/ H_2O : 70/30 (%v/v)	2.5–20% (w/v)	
41	Acrylonitrile copolymers with BuA, MA, MMA[44,45]		DMF	6–15 wt.%	

Source: Reproduced with permission from Ref. 9. Copyright © 2003 Elsevier Ltd.
Note: DMAc: dimethylacetamide; DMF: *N,N*-dimethylformamide; HFIP: hexafluoroisopropanol; IP: isopropanol; FA: formic acid; MEK: methylethylketone; ETOH: ethanol; HOAc: acetic acid; ACTN: acetone; CS_2: carbon disulfide; CSA: camphor sulfonic acid.

gel-spun polyethylene fibers, and aramids are used in composites of various polymers.[52] The improvement in modulus and strength achieved by the properties at the fiber/matrix interface and therefore dependent on the surface area of the interface. Nanofibers, with their very high specific surface area, should therefore present good composite characteristics.[53]

2.1.3 Biomedical Applications

Most of the human tissues and organs can be deposited in nanofibrous forms or structures, that is, bone, dentin, collagen, cartilage, and skin. Current research in electrospun polymer nanofibers has focused one of their major applications on bioengineering. There is a promising potential in various biomedical areas. Biomedical applications of nanofibers are medical prostheses, smart clothes, drug delivery carriers, wound dressings, cosmetic skin masks, and tissue scaffolds.

Some of the biomedical application of nanofibers is to cure for wound and burnings in human skin. It can be designed for especially hemostatic tools. Electrospun biodegradable polymers can be spun onto the wound skin. They form a thin web onto the skin. This web protects skin from microbes. Moreover, it helps to heal the wound quickly. Finally, it minimizes the possibility of scars. Electrospun nanofiber equipment used in wound healing is shown in Fig. 2.2.

Nanofibrous mats are being explored as biomedical grafts and wound dressings. It has been found that cells can adhere to and proliferate into the mats with a great deal of success. Also, because of the extremely small size of the nanofibers, the potential exists for layering of different polymers with specific functionalities.

Nanofibrous webs have the advantage due to their high surface-area-to-volume ratio of the nanofibers. Nanofibrous mats have been found to have high antibacterial effectiveness, which creates potential uses for them as high-performance filters, protective textiles, biomedical devices, wound dressings, hospital beddings, medical clothing for hospital staff, sports clothing, shoe linings, armbands, sleeping bags, and toys for children, as well as hygienic underwears and ladies tights.

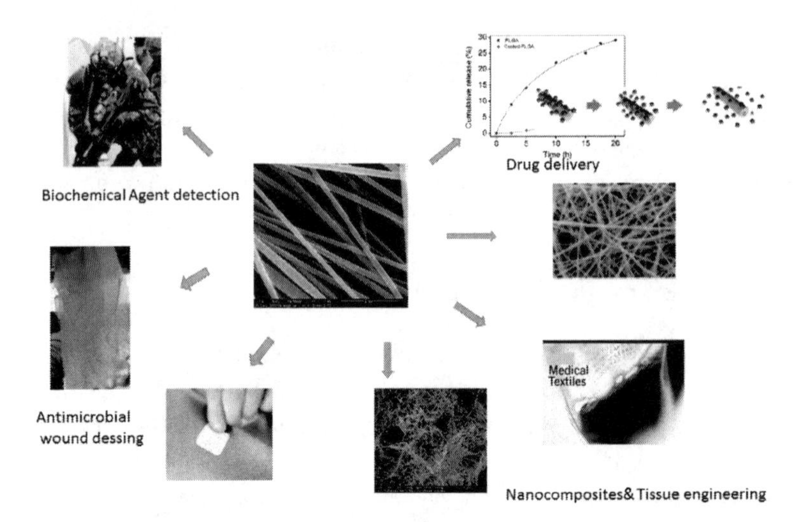

Figure 2.2 Application of electrospun nanofibers used in biochemical agent detecting, drug delivery, tissue engineering, wound covering and healing. Adapted from literature by A. Sezai Sarac.

Active ingredients, such as therapeutic compounds, antimicrobials, and antibiotics, were incorporated in electrospun nanofibers because the nanofiber webs have very strong efficacy for the drug due to their high surface-area-to-volume ratio and controlled release of the activity. The other ingredients of high importance in various biomedical fields, such as wound-dressing materials, body wall repairs, augmentation devices, tissue scaffolds, and antimicrobial filters, are silver (Ag) ions, silver compounds, and silver salts. For preventing bacterial attachment and biofilm formation on the surfaces, application of surface coatings or modification and/or alteration of the surface structure has been utilized.

Lee et al. prepared silver (Ag) nanoparticles in a polyacrylonitrile (PAN) nanofibrous film by sol–gel-derived electrospinning and subsequent chemical reduction for 30 min in hydrazine hydroxide (N_2H_5OH) aqueous solution.[54] Rheological properties of $AgNO_3$/PAN solutions are investigated as shown in Fig. 2.3. The viscosity of the $AgNO_3$/PAN precursor solutions increased initially up to the $AgNO_3$/PAN molar ratio of 0.05 and then reached a plateau region with further increasing the $AgNO_3$ content. The surface tension of the $AgNO_3$/PAN solution was not changed dramatically by the addition of $AgNO_3$. Antimicrobial properties of the $AgNO_3$/PAN precursor solution against gram-positive *Staphylococcus aureus* ATCC 6538 and gram-negative *Escherichia coli* ATCC 25922 were investigated. The formation of a clear zone suggested that the PAN solution containing Ag^+ ions was effective in the inhibition of bacterial growth. The Ag/PAN nanocomposite film, characterized by an X-ray diffraction (XRD), transmission electron microscopy (TEM), and ultraviolet-visible (UV-Vis) absorption spectrophotometer, revealed that crystallized cubic Ag particles with diameters of ~5.8 nm were dispersed homogeneously in PAN nanofibers.[54]

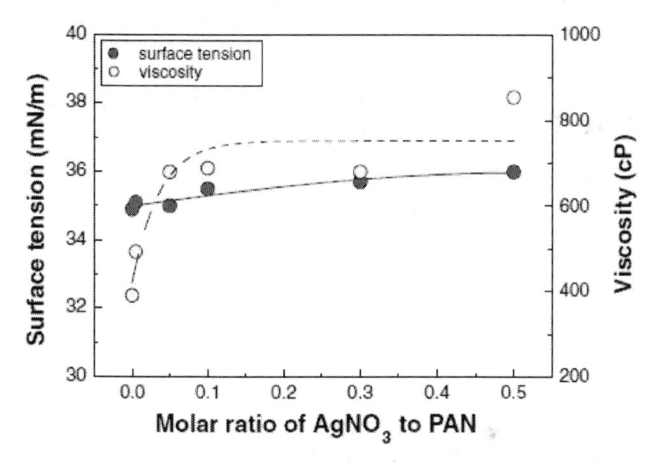

Figure 2.3 Variation of $AgNO_3$/PAN solution viscosity and surface tension as a function of the molar ratio of $AgNO_3$/PAN. Reprinted with permission from Ref. 54. Copyright © 2010, Springer Science+Business Media, LLC.

The Ag particle size determined from XRD measurements by the Scherrer equation was in the range of 5.8–5.1 nm (Fig. 2.4). TEM observations showed that Ag particles were dispersed homogeneously on and in PAN nanofibers (Fig. 2.4). A UV-Vis absorption band with a sharp maximum at $\lambda = 422$ nm, characteristic of bands of Ag nanoparticles, was observed.[54]

Antibacterial membranes with a multicomponent system containing Ag, AgBr, TiO_2, and hydroxyapatite as four active components were used to obtain more efficient antibacterial activity.[55] Additionally polyurethane nanofiber webs containing silver nanoparticles using the electrospinning technique were obtained with the stability of nanoparticles after washing cycles

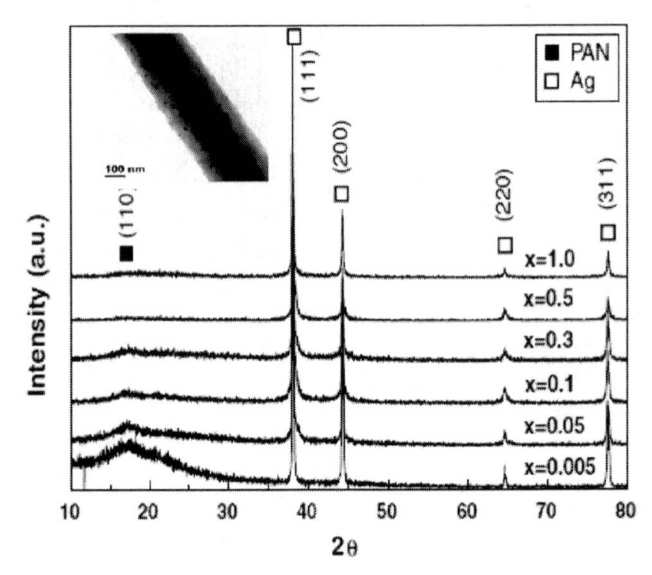

Figure 2.4 XRD patterns of Ag/PAN fibers after chemical reduction for 30 min in a N_2H_5OH aqueous solution. Note that *x* denotes the molar ratio of $AgNO_3$/PAN and (inset) TEM image of single PAN nanofibers embedded with Ag nanoparticles. Reprinted with permission from Ref. 54. Copyright © 2010, Springer Science+Business Media, LLC.

and the use of electrospun nanofiber mats in water purification was recommended. Electrospun nylon-6 nanofiber mats containing TiO_2 nanoparticles for water filter applications were also developed. The incorporation of TiO_2 nanoparticles in nylon-6 solution was found to improve the hydrophilicity, mechanical strength, and antimicrobial and UV protecting properties of electrospun polymeric webs. The UV-protective and antibacterial properties of electrospun TiO_2/PVA nanofiber webs as a multifunctional textile nanocomposite material was reported.[56] The TiO_2 particles content in the suspension were an 80%:20% anatase-to-rutile mixture and the average particle size was ~20 nm. Electrospinning solutions were prepared by dissolving the PVA polymer powder in distilled water at 80°C and stirring for six hours, followed by mixing with colloidal TiO_2 and stirring for an additional two hours. Scanning electron microscopy (SEM) images of an electrospun PVA fiber with and without TiO_2 are shown in Fig. 2.5. A TEM image and an energy-dispersive X-ray spectroscopy (EDX) spectrum of an electrospun TiO_2/PVA nanocomposite fiber are shown in Fig. 2.6.

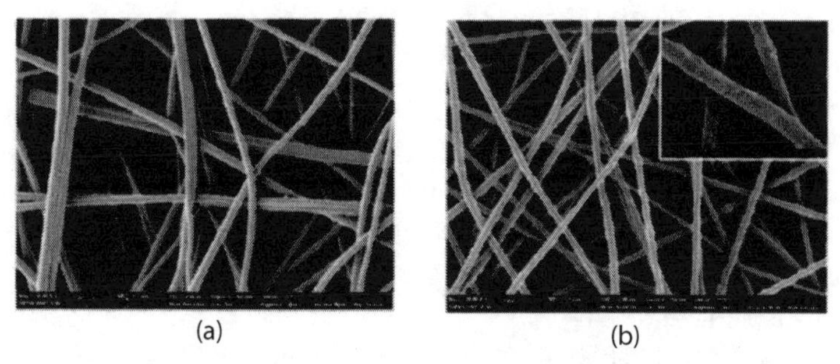

Figure 2.5 SEM images of (a) electrospun PVA nanofibers and (b) TiO_2/PVA nanocomposite fibers. Reprinted with permission from Ref. 56. Copyright © 2011 Wiley Periodicals, Inc.

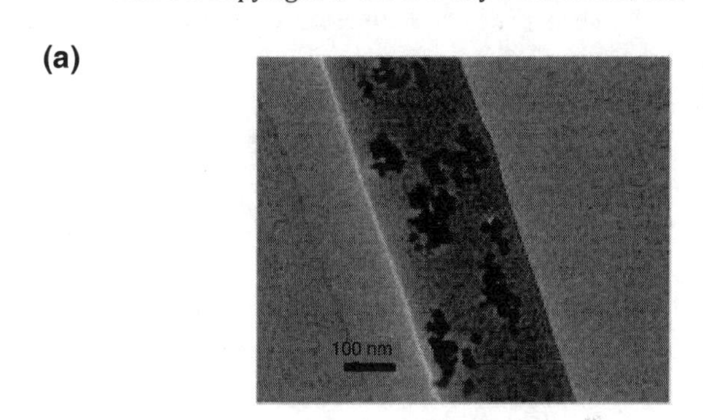

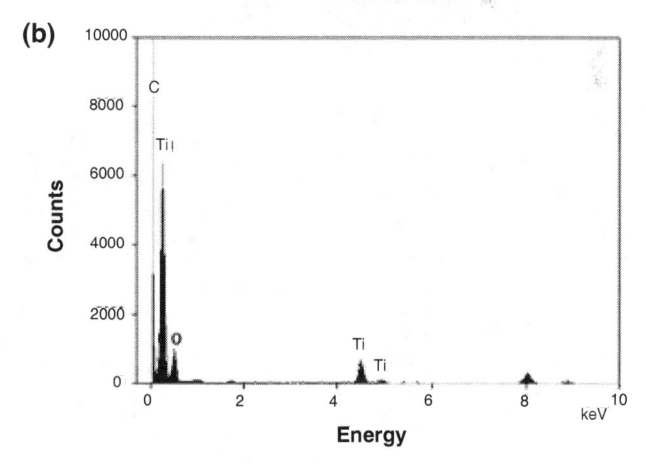

Figure 2.6 (a) TEM image of an electrospun TiO_2/PVA nanocomposite fiber and (b) EDX spectrum of an electrospun TiO_2/PVA nanocomposite fiber. Reprinted with permission from Ref. 56. Copyright © 2011 Wiley Periodicals, Inc.

Due to its biocompatibility and nontoxic nature, poly(*N*-vinyl pyrrolidone), PVP, a synthetic water-soluble polymer, finds applications in biomedical fields such as a blood plasma expander and a vitreous humor substitute. Since PVP is a water-soluble polymer, it has been used for controlled release of drugs by the oral route, also in pharmacy as a protective colloid, a viscosity-enhancing agent, a solubility promoter, a granulating, tableting agent, and a film-forming material.

The UV-cured film of *N*-vinyl pyrrolidone has been used as a potential bioadhesive wound-dressing matrix when blended with other polymeric materials. Skin covers and wound dressings made of PVA and PVP were produced with or without polysaccharides by the help of the gamma irradiation technique. Because of its biocompatibility and nontoxicity, PVP was also chosen as the base material for a drug-loading device. Drug release from such membrane was found to depend on the crosslinking density, composition, and membrane thickness. On the other hand, PVP has been used as a carrier in electrospun fiber applications.

A wound-healing substance was produced with PVP—water-soluble, biocompatible biopolymer—and the active ingredient (cetyltrimethyl ammonium bromide [CTAB] or cetylpyrridinum chloride [CPC]) as an organic material. It also has a potential drug release capacity due to its hydrophilic properties.

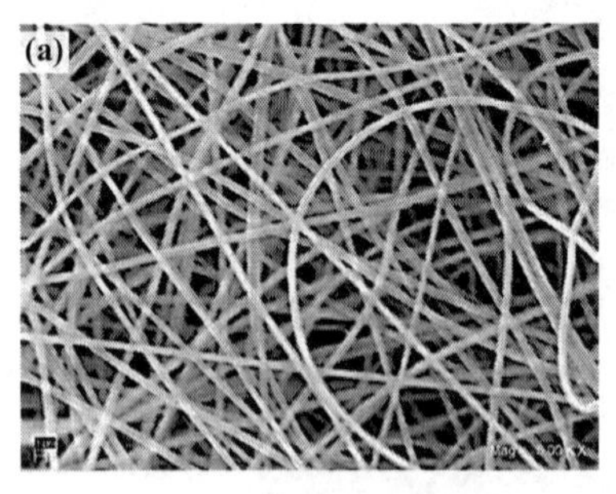

Figure 2.7 SEM image of electrospun samples (a) S1 and (b) S5 at a magnification of 5000x and 10000x using constant electrospinning parameters of 10 kV voltage, a needle–collector distance of 10 cm, and a flow rate of 1.0 mL/h. Reprinted with permission from Ref. 57. Copyright © 2014, SAGE Publications.

An antibacterial nanofibrous membrane was prepared by electrospinning of PVP/CTAB salt into nanofibrous webs. The individual features of PVP and the quaternary ammonium compound (CTAB or CPC) have combined to utilize them to develop an effective material for wound-healing applications. SEM micrographs of different samples containing different PVP/CTAB ratios have exhibited a smooth surface with a diameter range of 555.61 ± 86 nm (Table 2.2). The introduction of CTAB salt into the nanofibers affected the fiber morphology by a decrease in the average fiber diameter of PVP/CTAB samples ranging from ~500 nm to ~200 nm compared to that of the pristine PVP samples without salt (Figs. 2.7 and 2.8). As the quaternary ammonium group content of a polymer increases, because of the increased charge density of the polymeric solution needed for the electrospinning process, the average diameter size of the electrospun polymers decreases. In addition to providing antibacterial efficacy for electrospun mats, this is also another positive impact of the quaternary ammonium group on the polymers processed with electrospinning.[57]

The AFM micrographs of the control sample S1 and sample S5 having the highest CTAB content with 5.0% (w/v) are shown in Fig. 2.9. The pristine PVP nanofibers have a smooth surface and a homogenously dispersed structure with a surface roughness of 139.76 nm, calculated with Nanosurf software, while sample S5 demonstrates the presence of CTAB salt crystalline particles with a roughening effect on the fibers results in an increase in surface roughness by 180.96 nm.[57]

The nanofibrous web structure of the PVP membrane provides a large surface area to absorb water molecules, which potentially increases the drug dissolution rate.

Table 2.2 PVP and CTAB ratios (% w/v) of five different nanofibrous web samples (S)

Sample No.	PVP (% w/v)	CTAB (% w/v)	Fiber Diameter (nm)
S1 (control)	10.0	0.00	555 ± 86
S2	10.0	0.5	490 ± 12
S3	10.0	1.0	323 ± 19
S4	10.0	2.5	252 ± 12
S5	10.0	5.0	213 ± 93

Source: Reproduced from Ref. 57, Copyright © 2014, SAGE Publications.

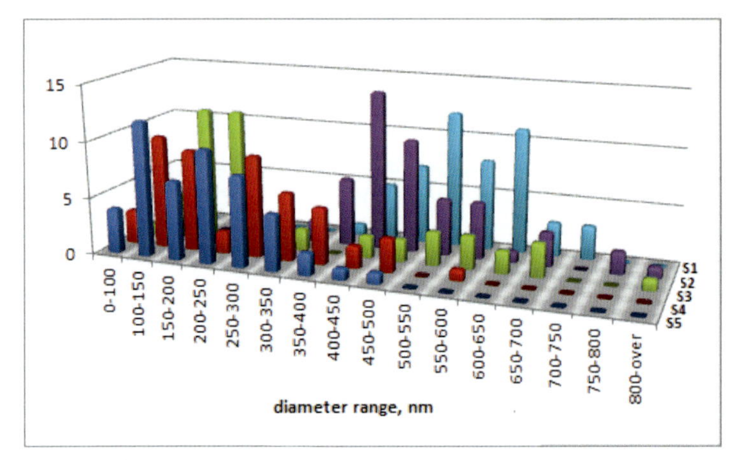

Figure 2.8 Fiber diameter distribution for the samples S1 through S5. Reprinted with permission from Ref. 57. Copyright © 2014, SAGE Publications.

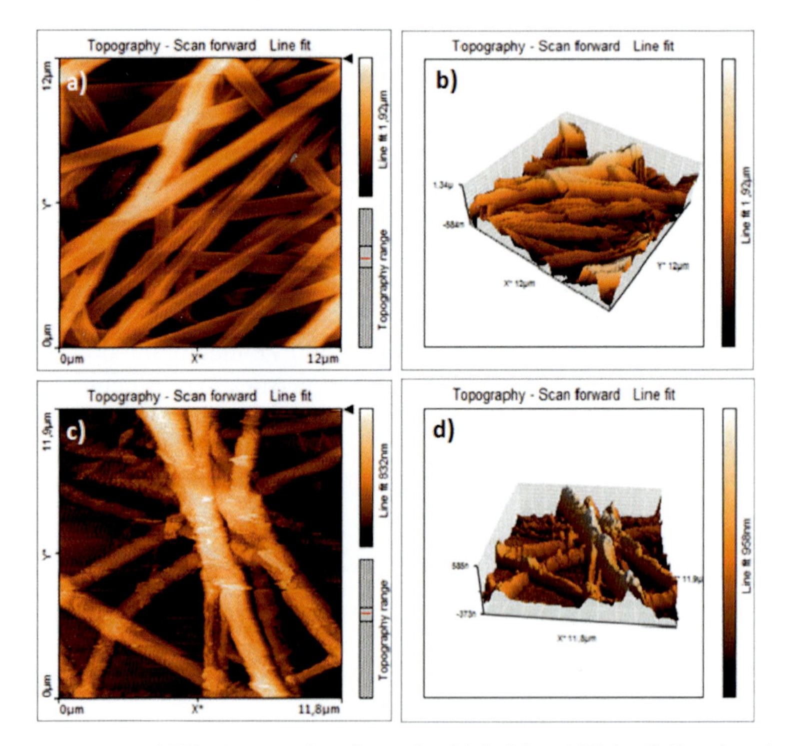

Figure 2.9 AFM micrographs of samples S1 (a, b) and S5 (c, d). Reprinted with permission from Ref. 57. Copyright © 2014, SAGE Publications.

Klebsiella pneumonia ATCC 4352, *E. coli* ATCC 10536, and *S. aureus* ATCC 6538 were the three different bacteria with gram-postive and gram-negative characteristics that were selected to determine the antimicrobial activity of nanofibrous webs made from PVP/CTAB salt. A detailed chemical characterization was performed on the produced nanofibrous webs by carrying out Fourier transform infrared spectroscopy (FTIR), ultraviolet visible (UV-Vis), and electrochemical impedance spectroscopy (EIS) measurements.[57]

2.1.4 Agricultural, Electrical, Optical, and Other Applications

Many synthetic and natural polymeric nanofibers can be prepared with various morphologies (e.g., cylinder shaped, foamed, wrinkled, and ribbon shaped). Different nanofillers (e.g., carbon nanoparticles, inorganic oxides, silicates, and carbon nanotubes) can be incorporated into polymeric nanofibers with the fillers' alignment along the fiber axes. The nonwoven/fabrics produced from electrospun nanofibers have the feature of controlling the pore sizes among nanofibers. Polymer nanofibers can be used as templates for the preparation of different nanofibers and/or nanotubes. Electrospun polymeric nanofibers have military and commercial interests, e.g., composites, filtration/separation, protective clothing, catalysis, agriculture, biomedical applications (e.g., tissue engineering and drug delivery), electronic applications (e.g., capacitors and transistors), and space applications as mentioned in previous sections.

Plants can be covered with a web that produced by electrospun nanofibers (Fig. 2.10). One of the functions of this web is protection against harmful insects and chemicals. It can be used as a greenhouse covering and in fertilizer applications, which are injected to the web before.

In military applications, protective clothes can hold the possibility of survival; enable long-term protection; stand heavy weather conditions; endure nuclear, chemical, and biological effects; and increase efficiency. Protective clothes now being used are made of heavy fabrics. The light and high porous fabrics that absorb the air and air vapor can easily react with chemical gases that cause the fabrics to decompose. Because of the higher surface area and low pore size of the nanofibers, the fabrics, produced by them, are suitable for protecting clothes. Additionally, neutralization can be provided with these fabrics. Also, they allow the clothes to breathe. Nanofibers that have high numbers of pores with small pore sizes provide high resistance to penetration of chemicals into the fabrics. Military applications are benefited from nanosensors for the work of finding traces, from nanoelectronics for various controls, and from nanocomposites for platforms that need lightness.

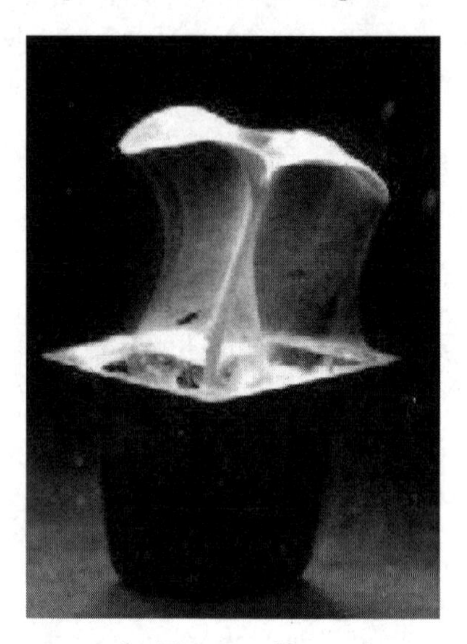

Figure 2.10 A plant covered with a nanofiber web.

Production of nanofibers that have the ability to transmit electricity creates many advantages. These nanofibers are used in the production of small electronic devices and in the fabrication of some machines. Because the surface area of the electrodes is proportional to the chemical reaction speed, electrospun nanofiber membranes are used appropriately in the production of improved high-performance batteries.

Nanofibers also used for sensors because their high specific area allows them to sorb and/or react rapidly with low levels of analytes in the air.[58,59] It is reasonable therefore to expect better performance from nanofiber sensors.

The other potential applications of nanofibers are wires, capacitors, transistors, and diodes for information technology and enzyme carriers. The application of nanofibers in textiles is not aimed only at specialized industries involved in technical textiles.[60]

2.2 Electrospun Ceramic Nanofibers and Their Applications

Ceramic nanofibers (e.g., SiO_2, Al_2O_3, ZnO, and TiO_2 nanofibers) can be prepared via electrospinning of precursor nanofibers followed by pyrolysis. For example, the precursor of TiO_2, titanium tetraisopropoxide, $Ti(OC_3H_7)_4$, can be coelectrospun with a matrix polymer of PVP using an ethanol/acetic acid mixture as the solvent.[61] Since titanium tetraisopropoxide can be easily hydrolyzed by the moisture in air, networks (gels) of TiO_2 nanofibers can be produced during or shortly after electrospinning. In another work carbon nanospheres were mixed with poly(vinylpyrrolidone)/ethanol solution and titanium tetraisopropoxide for electrospinning; and subsequent calcination of as-spun nanofibers led to thermal decomposition of carbon nanospheres, leaving behind pores in the TiO_2 nanofibers. The porous TiO_2 nanofibers have higher surface area and enhanced photocatalysis activity, compared to nonporous TiO_2 nanofibers (Fig. 2.11).[62]

Figure 2.11 SEM image of carbon nanospheres (left-above) and after (left-below) calcination of nanofibers prepared with carbon nanospheres, and XRD patterns of TiO_2 nanofibers (right) (a) and porous TiO_2 nanofibers (b). Reprinted from Ref. 62, Copyright 2012 Shanhu Liu et al.

The photocatalytic activity of these nanofibers was used to decompose the aqueous suspension of methylene blue (MB) under UV light at room temperature. The concentration of dye solution was measured with UV irradiation time by using a UV-Vis spectrometer. To monitor the change in concentration of MB, the absorbance spectra of MB were recorded at its maximum absorption against time (Fig 2.12).[62]

The organic components in the as-electrospun composite nanofibers can then be selectively removed via pyrolysis of the samples in air at an elevated temperature, resulting in the formation of TiO_2 nanofibers. By controlling the gelation of $Ti(OC_3H_7)_4$ aqueous solution (i.e., adjusting the pH value), TiO_2 nanofibers can also be electrospun without the presence of carrying polymer matrices.

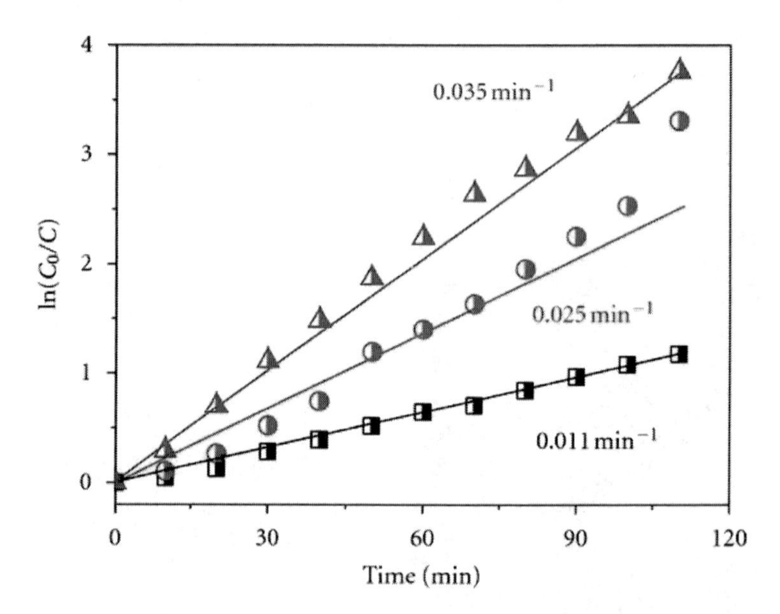

Figure 2.12 Liquid-phase photocatalytic degradation reaction kinetics of MB against TiO_2 nanofibers, TiO_2 P25 (*AEROXIDE* - commercial hydrophilic fumed titanium dioxide) and P-TiO_2 (porous) nanofibers. Reprinted from Ref. 62, Copyright 2012 Shanhu Liu et al.

2.3 Electrospun Metallic Nanofibers

In general, electrospun metallic nanofibers can be prepared using polymer and/or ceramic nanofibers as templates. The porous silver nanofibers can be prepared as follows: Polymers containing amidoxime functional groups ($-C(NH_2)=NOH$) on the surface possess high metal adsorption capacity. Amidoxime groups can be introduced onto the surface of electrospun PAN nanofibers through treatment with aqueous hydroxylamine (NH_2OH) solution. The nitrile ($-C{\equiv}N$) groups on the surface of PAN nanofibers react with NH_2OH and lead to the formation of amidoxime groups, which can be used for chelating silver ions upon immersion in $AgNO_3$ aqueous solution. The chelated silver ions can then be reduced by hydrazine (NH_2-NH_2) into elemental silver nanoparticles. Removal of PAN can be conducted via sintering of silver nanoparticles at lower temperature than the silver's melting point, since silver nanoparticles exhibit much lower melting points than bulk silver. Thiol groups can be introduced onto the surface of electrospun SiO_2 nanofibers through treatment with 3-mercaptopropyltrimethoxysilane. The thiol groups (R–SH) bind strongly with silver ions. The thiol-functionalized nanofibers can be immersed in $AgNO_3$ aqueous solution to adsorb silver ions. Reduction of silver ions to metallic silver nanoparticles can be achieved using hydrazine. Electrospun SiO_2 nanofibers will retain the fiber morphology during the sintering of silver nanoparticles. The SiO_2 templates can then be removed using hydrofluoric acid solution to obtain porous silver nanofibers. Porous silver nanotube networks with controlled nanostructures and superior surface-enhanced Raman scattering (SERS) activity can be utilized for in situ Raman detection of minerals, microbes, and biomarkers.*

2.4 Electrospun Carbon/Graphite Nanofibers

Carbon/graphite nanofibers (Fig. 2.14) are made by carbonization/graphitization of their precursors of electrospun polymer nanofibers (e.g., PAN nanofibers). Two types of carbon/graphite nanofibers can be developed, (type 1) continuous, nanoscaled carbon fibers with superior mechanical strength and (type 2) highly graphitic, porous graphite nanofibers with specific surface areas of up to 2500 m^2/g.

---

*Excerpt reprinted with permission from Ref. 63, Copyright Dr. Hao Fong.

Continuous nanoscaled carbon fibers can be developed by stabilization and carbonization of highly aligned and extensively stretched electrospun PAN copolymer nanofiber precursors under optimal tension. These carbon fibers with diameters being tens of nanometers possess a better mechanical strength, which is not easy to be achieved through conventional approaches. The reason for this is that the diameter of the fiber will be 100 times smaller than that of conventional counterparts, due to which the fibers will possess a high degree of macromolecular orientation and a significantly reduced amount of structural imperfections. The small fiber diameter will also prevent the formation of structural inhomogeneity, that is, sheath–core structures, during stabilization and carbonization.[63]

> Small and uniform gold nanoparticles (AuNPs) embedded in polyacrylonitrile nanofibers (PANFs) were synthesized via a combination of electrospinning and in situ reduction. With the conversion from the amorphous structures of PANFs to graphene layered structures of carbon nanofibers (CNFs), the AuNPs can migrate from the interior of PANFs to the external surfaces of CNFs. The migration of AuNPs through the nanofiber matrix is strongly dependent on the graphitization temperature and heating rates. Three different heating rates of 2, 5, and 10°C min^{-1} and graphitization temperatures of 600, 800, and 1000°C were employed to investigate the migration and the exposed density of AuNPs on the CNFs* (Fig 2.13).

Acrylonitrile copolymers and PAN nanofibers were prepared by electrospinning from *N,N*-dimethylformamide (DMF) solution, on which structure and electromagnetic properties were studied. Structural analysis was performed on carbon nanofibers, which were prepared from PAN/DMF solution by carbonization at 750°C followed by 1100°C.[64,65] The obtained carbon nanofibers had an average diameter of 110 nm, an interlayer spacing d 002 of 0.368 nm, and a Raman band intensity ratio I D /I G of 0.93.

SEM and TEM observations indicated that fibers exhibited a skin–core heterogeneity in the skin, carbon layers being oriented predominantly parallel to the fiber surface. PAN-based carbon nanofiber bundles prepared from 10 wt.% PAN/DMF solution added with 5 wt.% acetone and 0.01 wt.% dodecylethyldimethylammonium bromide and collected on the rim of the rotating disc covered with Al

*Excerpt reprinted with permission from Ref. 64, Copyright 2014, Royal Society of Chemistry.

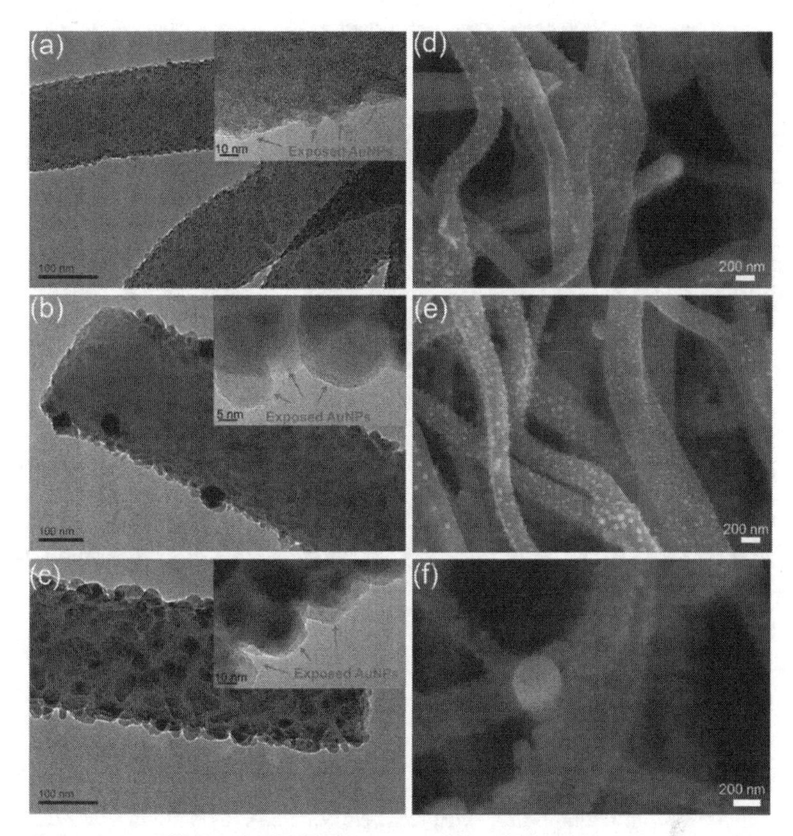

Figure 2.13 TEM and FE-TEM images of Au–CNFs with various graphitization temperatures at (a and d) 600°C, (b and e) 800°C and (c and f) 1000°C, respectively. Reprinted with permission from Ref. 64. Copyright © 2014, Royal Society of Chemistry.

foil were subjected to heat treatment at 1400°C, 1800°C, and 2200°C for one hour.[66] The diameter of nanofibers composing the bundles was approximately 330 nm for as-spun, 250 nm for 1000°C-treated, and 220 nm for 1800°C-treated fibers. TEM images are shown in Fig. 2.15 on 1000°C-treated and 2200°C-treated nanofibers the latter having d_{002} of 0.344 nm and I_D/I_G larger than 1.0.

To obtain better alignment of basic structural units of hexagonal carbon layers along the fiber axis, multiwalled carbon nanotubes (MWCNTs) were embedded into electrospun PAN-based carbon nanofibers, although the improvement was observed around MWCNTs.[67] TEM observation on MWCNT-embedded PAN-based

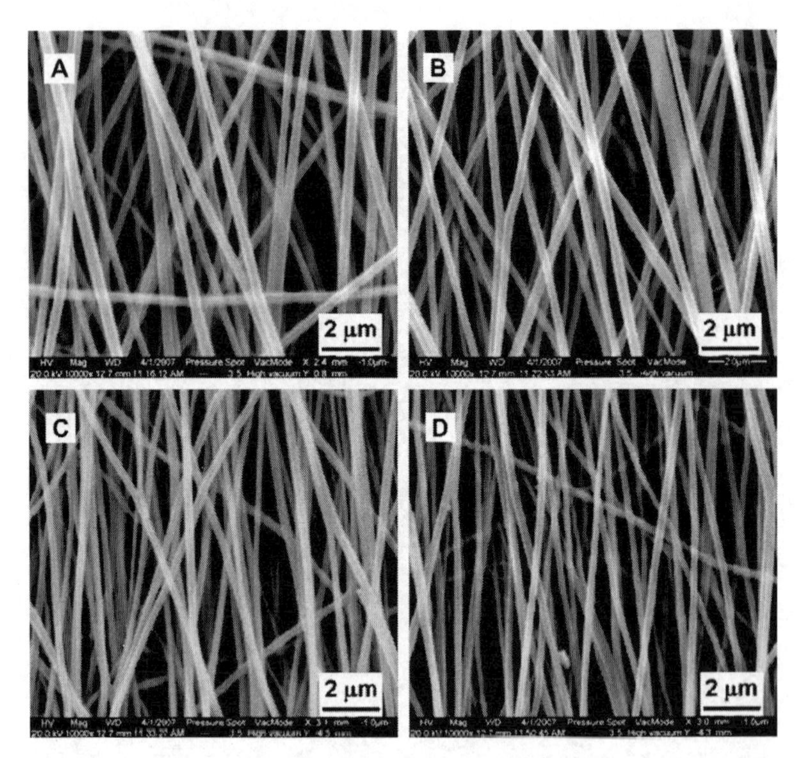

Figure 2.14 Representative morphologies of (A) as-electrospun PAN nanofiber bundle, (B) stabilized PAN nanofiber bundle, (C) low temperature (1000°C) carbonized PAN nanofiber bundle, and (D) high temperature (2200°C) carbonized PAN nanofiber bundle. Reprinted with permission from Ref. 66. Copyright © 2009 Elsevier Ltd.

nanofibers by in situ heating up to 750°C showed only a local orientation of carbon layers.[68] On a PAN-based nanofiber web after the activation by steam at 800°C, adsorption behavior of benzene vapor was studied at a temperature of 343–423 K under a pressure of up to 4.0 kPa, confirming high adsorption in comparison to activated carbon fibers A-10.[69] PAN electrospinnability and commercial viability were reviewed.[70] Core–shell polymeric nanofibers were electrospun through a doubled capillary, PAN/DMF solution in the outer capillary and poly(methyl methacrylate) (PMMA) in the inner capillary, and converted to hollow carbon nanofibers by carbonization up to 1100°C.[71] Similar hollow nanofibers were synthesized by electrospinning of emulsion-like DMF solution of PAN

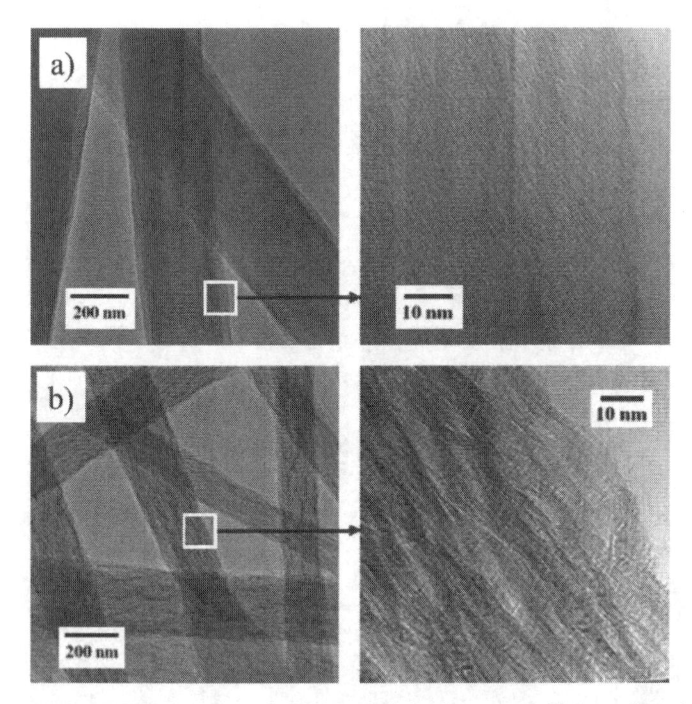

Figure 2.15 TEM images of PAN-based carbon nanofibers after heat treatment at 1000°C (a) and 2200°C (b). Reprinted with permission from Ref. 66. Copyright © 2009 Elsevier Ltd.

and PMMA in different ratios through a single capillary of 0.5 mm diameter, followed by carbonization at 1000°C and heat treatment up to 2800°C,[72] as shown in Fig. 2.16. By changing the PAN/PMMA ratio, the mesopore volume could be controlled, that is, the mesopore volume changed from 0.18 cm^3g^{-1} for a 9/1 ratio to 0.47 cm^3g^{-1} for a 5/5 ratio, although the micropore volume did not change much and became 0.34 cm^3g^{-1}.[72]

High-temperature treatment improves conductivity of carbon fibers, not only due to the development of a graphitic structure, but also due to exclusion of foreign atoms in carbon materials.

To increase electrical conductivity of electrospun carbon nanofibers, embedding of MWCNTs was performed in relation to electric double-layer capacitor (EDLC) application.[73] Embedding of 0.8 wt.% MWCNTs led to an increase in the electrical conductivity of the webs from 0.86 to 5.32 S/cm, accompanied by an increase in EDLC capacitance from 170 to 310 $F.g^{-1}$ in 1 M H_2SO_4 aqueous electrolyte.[74]

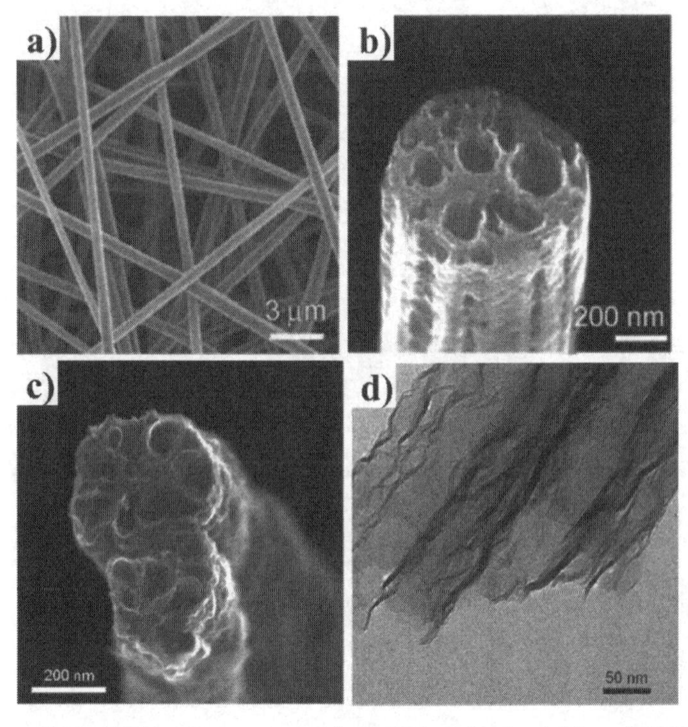

Figure 2.16 SEM images as-spun (a), 1000°C-carbonized (b), and 2800°C-treated (c) fibers and TEM image of 2800°C-treated fiber (d) prepared from the mixture of PAN/PMMA:5/5. Reprinted with permission from Ref. 72. Copyright © 2007 WILEY-VCH Verlag GmbH & Co. KGaA, Weinheim.

Chapter 3

Conjugated Polymers

3.1 Conjugated Polymers

Nanotechnology and application as conjugated polymeric materials in nanofibers, nanocoreshell, nanocomposite, nanolithography, and semiconductive products provides a wide range of implementation, such as sensors, fuel cell, and organic light-emitting diode.[75]

3.1.1 Conductivity of Conjugated Polymers

Conducting polymers (CPs) behave like semiconductors in terms of conductivity but normally when they are synthesized they indicate low conductivity because of being neutral. To behave as metallic conductors, they have to be doped by a dopant during the doping process. Doping is a process of addition of electrons (reduction) or removal of electrons (oxidation) from the polymer chain. The charged species formed are able to move along the carbon chain (delocalization), allowing electron transport and thus giving an electronically conductive material. Once doping has occurred, the charge-carrying species in the delocalized π system have the mobility

Nanofibers of Conjugated Polymers
A. Sezai Sarac

ISBN 978-981-4613-51-4 (Hardcover), 978-981-4613-52-1 (eBook)
www.panstanford.com

to move along the backbone chain. From a macroscopic perspective, conduction through a CP takes place by charge hopping both along the polymer chains and also between the macromolecules that make up individual fibers and between the fibers themselves. Doping (*p* or *n*) generates charge carriers that move in an electric field. Positive charges (holes) and negative charges (electrons) move to opposite electrodes. This provides movement of charge for electrical conductivity.[76]

Generally, organic CPs have backbones that contain alternating double and single bonds. These polymers possess semiconductor characteristics. In semiconductors, there is a small energy gap between the highest occupied molecular orbital (HOMO) or valence band (VB) and the lowest unoccupied molecular orbital (LUMO) or VB. So electrons can be excited either thermally or electrically over the gap where they are free to delocalize over the LUMO level or conduction band (CB). If there are enough small band gaps, a large delocalized band appears over the lattice, the electrons flow in the CB, and/or the vacant holes of positive charge flow in the VB and a current flow with electrons. The mechanism of conduction in conductive polymers involves the concept of solitons, polarons, and bipolarons. Influencing factors for conductivity are the polaron length, the conjugation length, the overall chain length, and the charge transfer to adjacent molecules. These factors are explained by models based on hopping between localized states assisted by lattice vibrations, intersoliton hopping, variable range hopping in three dimensions, intrachain hopping of bipolarons, and charging energy–limited tunneling between conducting domains.[1]

3.1.2 Electronic Conduction

The electronic and optical properties of π-conjugated polymers result from a number of states around the highest occupied and the lowest unoccupied levels. The highest occupied band, which originates from the HOMO of each monomer unit, is referred to as the VB. The corresponding lowest unoccupied band, originating from the LUMO of the monomer, is the CB (Fig. 3.1).

The energy distance between VB and CB is defined as the band gap (Eg) and refers to the onset energy of the π–π^* transition in neutral conjugated polymers. The E_g of conjugated polymers can be approximated from the onset of the π–π^* transition in the UV-Vis spectrum. Conjugated polymers behave as semiconductors in their neutral state. However, upon oxidation (*p*-doping) or reduction (*n*-doping), the interband transitions between the VB and the CB can decrease the effective band gap, thereby resulting in the formation of charge carriers along the polymer.[77]

Band theory application to conjugated polymers was focused on polyacetylene; in the neutral state the two resonance forms of polyacetylene are degenerate and on oxidation they lead to the formation of solitons. The localized electronic state associated with the soliton is a nonbonding state at an energy lying in the middle of the π–π^* gap, between the bonding and antibonding levels of the polymer chain. The soliton is a topological as well as a mobile defect because of the translational symmetry of the chain.[77]

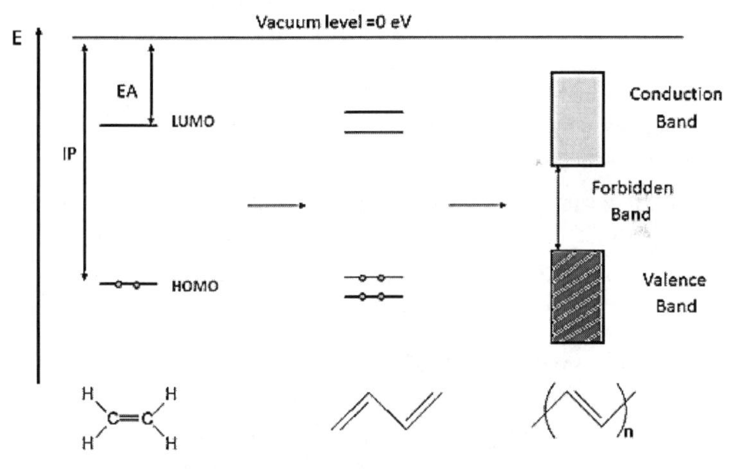

Figure 3.1 Schematic representation of ethylene, a dimer of acetylene, and polacetylene band gap (one electron structure) IP: ionization potential; EA: electron affinity.

Conjugated polymers have a conjugated π system and π bands. As a result, they have a low ionization potential (usually lower than ~6 eV) and/or a high electron affinity (lower than ~2 eV).

They can be easily oxidized by electron-accepting molecules (I_2, AsF_5, SbF_5) and/or easily reduced by electron donors (alkali metals: Li, Na, K).

Charge transfer between the polymer chain and dopant molecules is easy after doping neutral conjugated molecules: *n*-doping corresponds to reduction (addition of electron), and *p*-doping corresponds to oxidation (removal of electron).

The soliton model was first proposed for degenerated CPs (polyacetylene [PA] in particular) and it was noted for its extremely one-dimensional character, each soliton being confined to one polymer chain. Thus, there was no conduction via interchain hopping. Furthermore, solitons are very susceptible to disorder and any defect such as impurities, twists, chain ends, or crosslinks will localize them.

The application of an oxidizing potential to aromatic polymers with nondegenerate ground states destabilizes the VB, raising the energy of the orbital to a region between the VB and the CB. Removal of an electron from the destabilized orbital results in a radical cation or polaron. Further oxidation results in the formation of dications or bipolarons dispersed over a number of rings. These radical cations are the charge carriers responsible for conductivity in conjugated polymers. Because of the nondegenerate energy transitions of conjugated polymers (excluding PA), structural changes result and backbone will localize them.*

An ionization potential versus chain length graph (Fig. 3.2) indicates that it is easier to remove an electron from a long oligomer (oxidize) than the monomer itself.

3.2 The Molecular Structure of PEDOT

Poly(3,4-ethylenedioxythiophene) (PEDOT) lacked the presence of undesired couplings within the polymer backbone. PEDOT was initially found to be an insoluble polymer exhibiting interesting properties. In addition to very high conductivity (ca. 200 S/cm), PEDOT was found to be almost transparent in the thin, oxidized state. The ionization potential versus chain length (PEDOT) (n = 1 monomer; n = large polymer between oligomeric species) dependency is

*Excerpt reprinted with permission from Ref. 77, Copyright 2014 IJTEEE.

shown in Fig. 3.2 (where NDB = number of bonds in the conjugated pathway).[78]

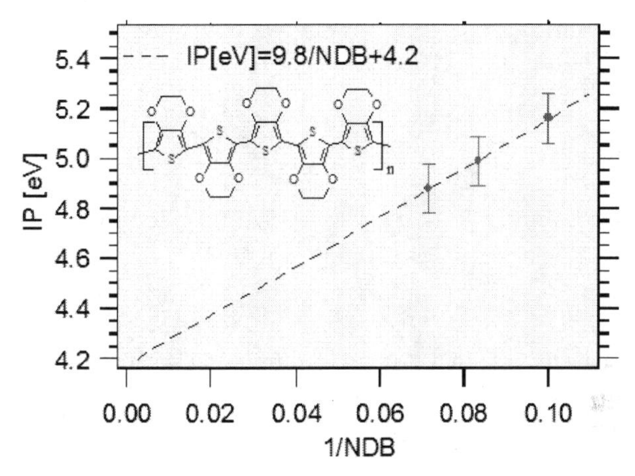

Figure 3.2 Ionization potential vs. chain length of poly(3,4-ethylenedioxythiophene (PEDOT) (n = 1 monomer; n = large polymer between oligomeric species; NDB = number of bonds in the conjugated pathway). Reprinted with permission from Ref. 78. Copyright 2003, AIP Publishing LLC.

3.2.1 Doping in Poly(*p*-Phenylene Vinylene) and Poly(3,4-Ethylenedioxythiophene)

Regarding the nature of the excited states that are created when photons are absorbed, the experimental results are presented in Fig. 3.3. The figure shows absorption spectra in the Vis range, which were obtained for a series of oligomers of poly(*p*-phenylene vinylene) (PPV). The four presented absorption bands all have a similarly fine structure and are shifted against each other; the higher the number of monomers, the lower the frequency. A first conclusion can be drawn from the invariance of the profile of the bands. Since it is independent of the number of coupled monomers it must be a property of the monomer itself.

Figure 3.4 presents its chemical structure on the left-hand side and on the right-hand side a sketch of the distribution of the π electrons. One bond of each double bonds is set up by two σ electrons;

the other is formed from overlapping atomic *pz* orbitals oriented perpendicularly to the *xy* plane of the planar molecular skeleton.

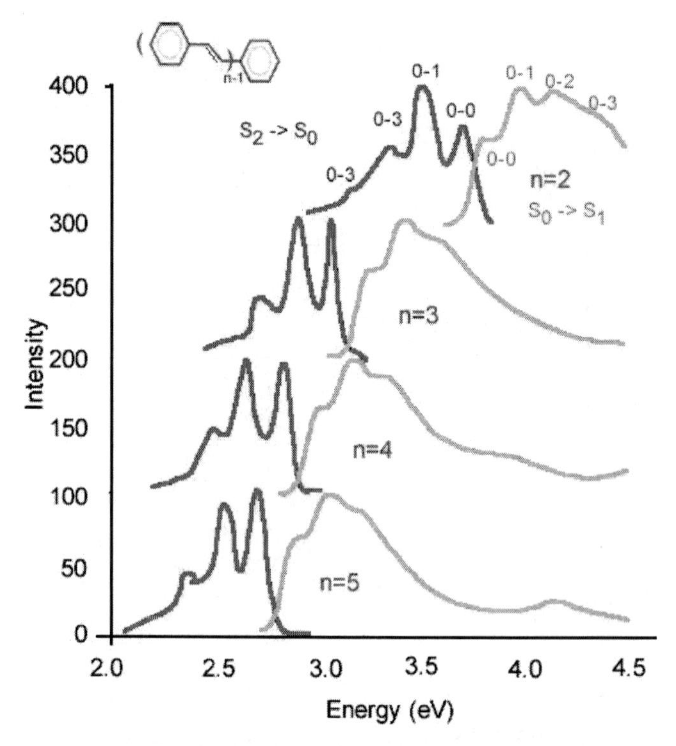

Figure 3.3 Experimental absorption and emission spectra of oligophenylenevinylenes at 77K in inert PMMA matrix. A shift to lower energies with increasing length of the molecule is observed. [79].

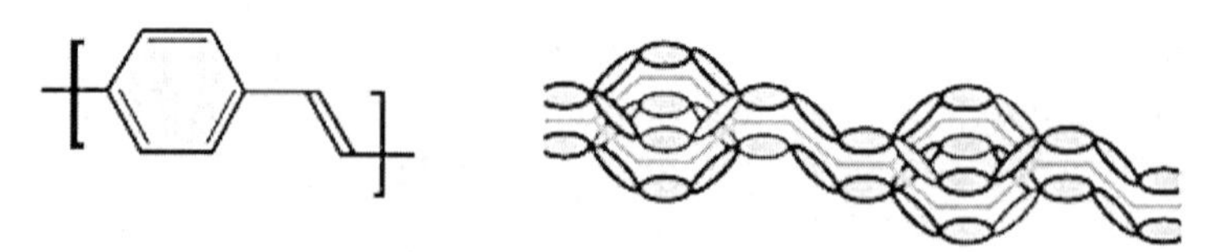

Figure 3.4 (Left) Structure unit of poly(*p*-phenylene vinylene)(PPV). (Right) Clouds of π electrons in a unit, placed above and below the plane of the CC backbone. The lines connecting the carbon atoms represent the σ bonds.

The **conjugation**, that is, the resonance interaction between the π bonds, results in delocalized π electron states. In the case of PPV these states are occupied by the eight π electrons. There is a highest occupied molecular π orbital, and there is a gap by extending up to next level, the lowest unoccupied molecular π orbital.

For conjugated polymers the gap energy is in the range of 1.5 eV to 3 eV, that is, in the range of Vis light and the near infrared (NIR), similar to inorganic semiconductors. In the ground state of a multielectron system with an even number of electrons, the spins compensate each other so that a singulet state results. Polymer chain defects and charge transport in conjugated polymers can be represented as in Fig. 3.5 with a schematic energy diagram for positive polarons without an electric field and charge transport under applied an electric field.

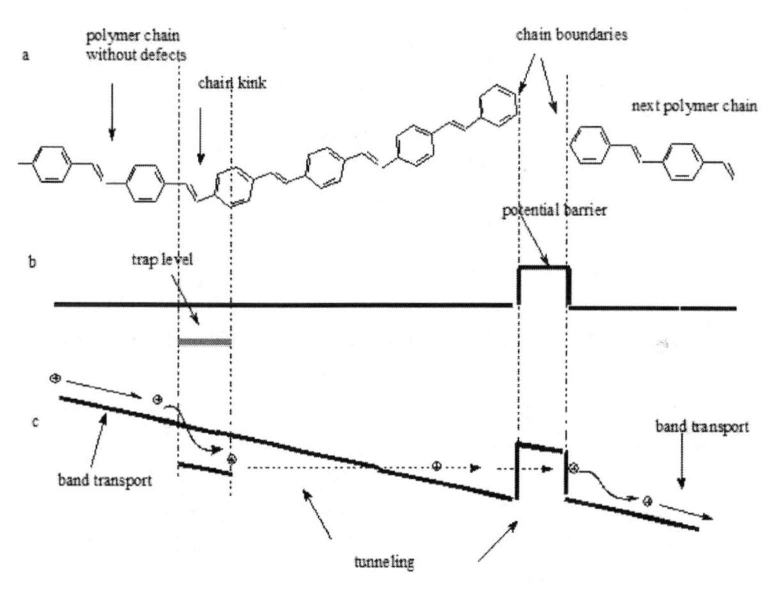

Figure 3.5 Polymer chain with characteristic defects (a); schematic energy diagram for positive polarons without an electric field (b); and charge transport under an applied electric field (c).

Electrochemical doping: The doping charge comes from the electrode and the ions of the salt included in the electrochemical bath play the role of the counterions (Fig 3.6).

Chemical doping: The doping charge (electron or hole) on the conjugated molecules or polymers comes from another chemical species (atom or molecule). The chemical species become the counterions of the polarons created on the conjugated materials.

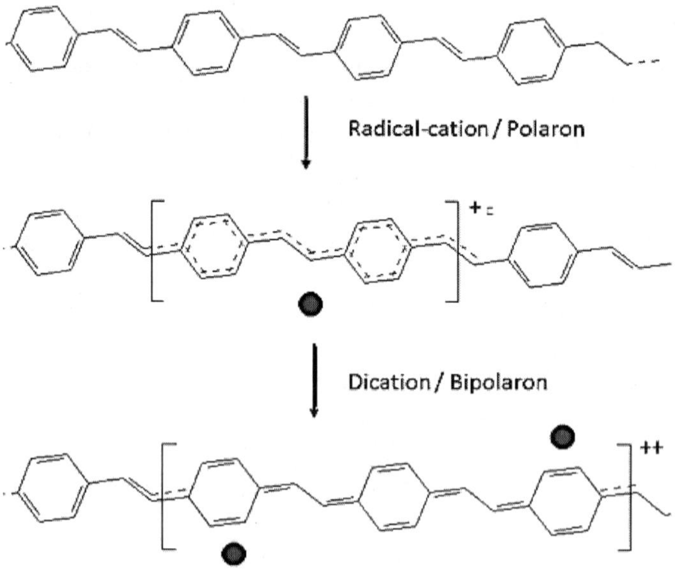

Figure 3.6 Polaron and bipolaron formation in PPV.

A strong **electron donor** (reducing agent) can be used to dope negatively, a neutral conjugated material or to undope a positively doped material, for example, tetrakis(dimethylamino)ethylene (TDAE) and alkali metals (Li, Na, K, etc.).

A strong **electron acceptor** (oxidizing agent) can be used to dope positively, a neutral conjugated material or to undope a negatively doped material, for example, nitrosonium tetrafluoroborate ($NOBF_4$) and halogen gases (I_2, etc.).

Examples of dedoping and doping: An effective dedoping method for organic semiconductors using TDAE was demonstrated using poly(3-hexylthiophene) (P3HT) thin-film transistors.

Dedoping: PEDOT–polystyrene sulfonic acid (PSS) (*p*-doped) is exposed to a vapor of TDAE and undergoes dedoping. Peaks in the infrared (IR) region disappear, and a peak in the Vis region appears.

The reduction order is the following:

Bipolaron → Polaron → Neutral

UV-Vis-NIR absorbance spectroscopic investigation of the PEDOT-PSS system indicated that pristine PEDOT-PSS films exhibit a large absorption band covering the beginning of the IR domain, from 1250 nm to higher wavelengths. This band is attributed to bipolarons, that is, the dicationic form. These dicationic charges usually result from the combination of polarons (RCs). UV-Vis-NIR absorbance spectroscopy is used to determine the oxidation level of PEDOT-PSS thin films treated with aqueous and organic reducing agents (Fig. 3.7). After reduction with reducing agents (Na_2SO_3 and $Na_2S_2O_3$), the absorption band around 900 nm attributed to the RC increased, while a concomitant decrease of the dicationic contribution was observed (above 1250 nm). Further reduction, obtained with strong reducing agents like TDAE and $NaBH_4$, led to the onset of a new band around 660 nm, corresponding to polymeric neutral chains (NCs). The reduction level of the PEDOT-PSS measured by UV-Vis-NIR spectrometry increases with the redox potential of the reducing agent.[80]

The presence of midgap levels in doped (oxidized) polymers deeply influences their spectroscopic properties that become dominated by two subgap (low-energy) transitions from the VB to the localized bipolaronic states (E_{b1} and E_{b2}, Figs. 3.8 and 3.9). This behavior is the origin of the electrochromism in conjugated polymers since it results in a huge change in their absorption spectrum. The main optical absorption band in the reduced neutral conjugated polymers is determined by a π–π^* transition through the band gap, hence falling in the UV-Vis region, while the low-energy bipolaronic transitions give absorption bands at lower energies. Thin films of polymers with $E_g > 3$ eV are colorless in the undoped states, while their doped form is colored because of absorption in the Vis region. In contrast, those with lower band gaps ($E_g \approx 1.5$ eV) show colored reduced forms, but possess faintly colored or colorless doped forms, as an effect of absorption bands shifted to the NIR region with just small residual absorption in the Vis region. PEDOT possesses a stable and highly transmissive sky-blue oxidized state, the absorption band of reduced PEDOT lies in the NIR region of the Vis spectrum, giving its distinctive dark-blue color, and its doped form (very light blue) has minimal tailing in the Vis region from its charged carrier transitions.[81]

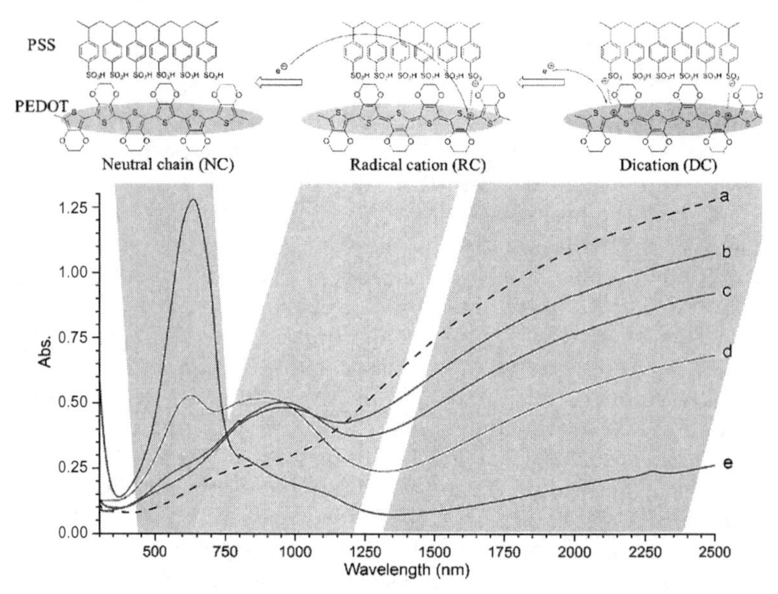

Figure 3.7 (Top) Chemical structures of a PEDOT-PSS: (left) a neutral chain, (center) an RC charge carrier, and (right) a dication charge carrier. (Bottom) Absorbance spectra of (a) pristine PEDOT-PSS and thin films treated with (b) $Na_2S_2O_3$, (c) Na_2SO_3, (d) $NaBH_4$, and (e) TDAE. Reprinted with permission from Ref. 80. Copyright © 2013, Royal Society of Chemistry.

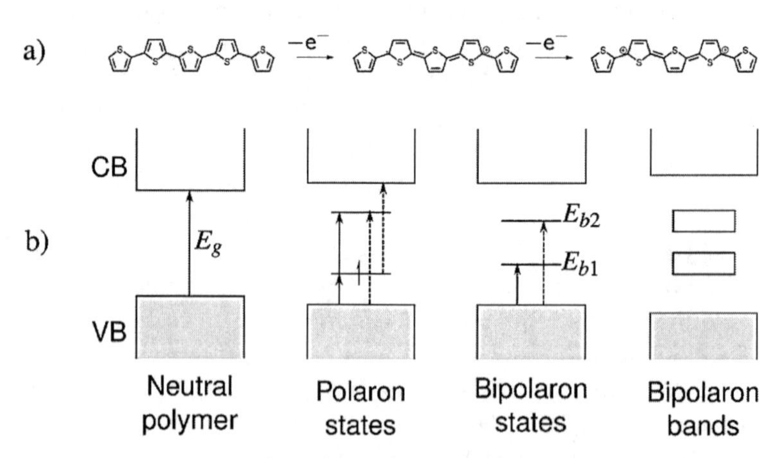

Figure 3.8 Transitions from the VB to the localized bipolaronic states (E_{b1} and E_{b2}).

PEDOT-PSS can be oxidized by a surface reaction with the strong electron donor TDAE.[82]

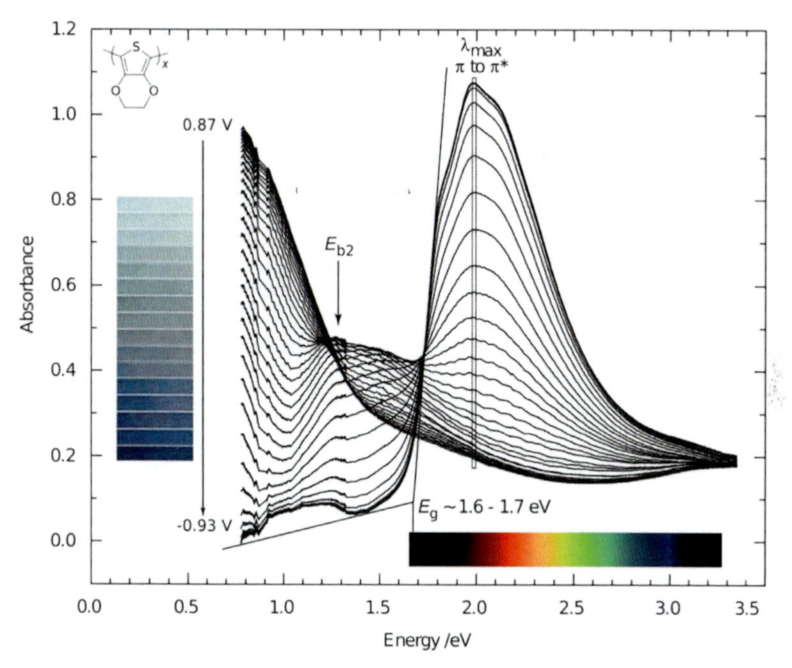

Figure 3.9 Spectroelectrogram of poly[3,4-(ethylenedioxy)thiophene] (PEDOT) film on ITO/glass. The film was electrochemically deposited from 0.3 M EDOT in 0.1 M Bu_4ClO_4/PC and switched in 0.1 M Bu_4ClO_4/ACN. Reprinted with permission from Ref. 81. Copyright © 2014, Royal Society of Chemistry.

The two electrons transferred from TDAE to PEDOT-PSS are expected to undope the conjugated polymer chains. Since TDAE diffuses into PEDOT-PSS, long exposures to the electron donor induce changes in the optical properties of the polymer film. Optical absorption experiments on 200 nm thick PEDOT-PSS films coated onto a transparent polyethylene terephthalate (PET) substrate. The polymer film was exposed to the TDAE vapor in an inert nitrogen atmosphere and shows the difference in absorption spectrum between a film exposed to TDAE and the pristine PEDOT-PSS layer (Figs. 3.10 and 3.11). The modification of the optical properties and the sheet resistance of the polymer layer were recorded versus exposure time. The two absorption features at 550 nm and

900 nm increased upon TDAE exposure, while the background in the NIR region characteristic of a doped conjugated polymer decreased. Those three features are attributed to the undoping of the positively charged PEDOT chains due to the electron transfer from TDAE. The modification in the optical spectrum of PEDOT-PSS upon TDAE exposure is due to the transformation of positive bipolarons (absorption in NIR) to polarons (absorption at ~900 nm) or neutral segments (absorption at 550 nm) of the PEDOT chains via electron transfer from TDAE. The band diagram of the two devices is represented in Fig. 3.10. The interface dipole due to the adsorbed TDAE molecules results in a vacuum level shift at the interface. The work function of the TDAE modified PEDOT surface is lowered to 3.9 eV, which is expected to significantly improve electron injection from this electrode.[82]

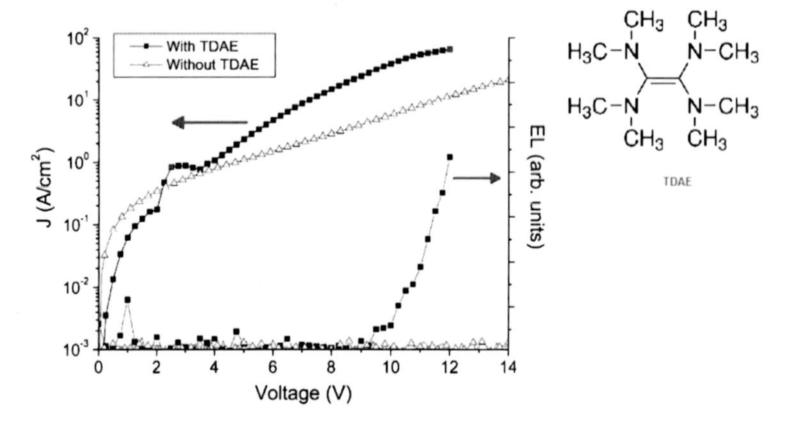

Figure 3.10 J-V behavior of all-plastic light emitting diode. Triangles and squares are from the hole-only reference device and the all-plastic light emitting diode with TDAE modified PEDOT-PSS respectively. Reprinted with permission from Ref. 82, Copyright © 2006 Elsevier B.V. All rights reserved.

PEDOT-C14 (neutral) is exposed to $NOBF_4$ and undergoes doping. The first peak in the IR region appears at about (800–1000 nm corresponding to polaron), and then a second broad band at higher wave wavelength disappears. The conductivity increases.

$$\text{PEDOT} - e \rightarrow [\text{PEDOT}]^{+}A^{-} - e \rightarrow [\text{PEDOT}]^{2+}(A^{-})_2$$

Neutral→ Polaron → Bipolaron

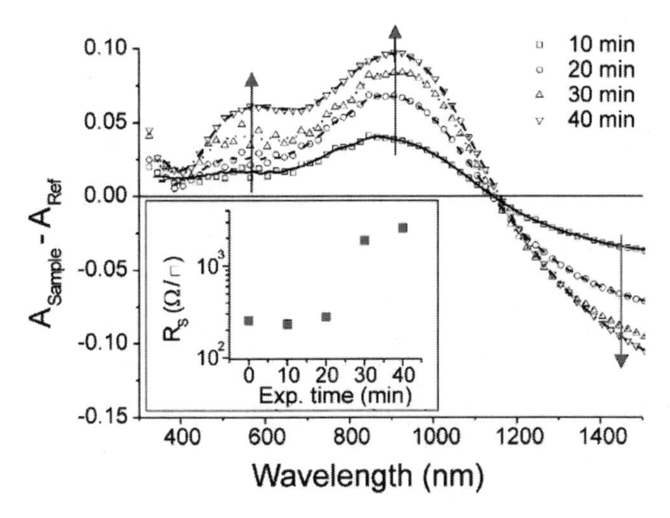

Figure 3.11 Absorbance changes with time. Reprinted with permission from Ref. 82, Copyright © 2006 Elsevier B.V. All rights reserved.

The application of an oxidizing potential to aromatic polymers with nondegenerate ground states destabilizes the VB, raising the energy of the orbital to a region between the VB and the CB. Removal of an electron from the destabilized orbital results in an RC forms a polaron structure. Further oxidation results in the formation of dications or bipolarons dispersed over a number of rings. These RCs are the charge carriers responsible for conductivity in conjugated polymers. Due to the nondegenerate energy transitions of conjugated polymers (excluding PA), structural changes result and are based on the widely accepted mechanism for polypyrrole (PPy). This mechanism has been supported by electron paramagnetic resonance (EPR) measurements, showing that neutral and heavily doped polymers possess no unpaired electrons, while lightly doped polymers display an EPR signal.*

Another approach is on the formation of π dimers instead of bipolarons during the oxidative doping of conjugated polymers. According to this concept polarons from separate polymer chains interact, forming an EPR-inactive diamagnetic species; this has been demonstrated in studies of thiophene-based oligomers. Despite the fact scientists have been able to interpret the band structure of conjugated polymers to tune their electrical, optical, and electrooptical properties.[84]

*Excerpt reprinted with permission from Ref. 83, Copyright Anwar-ul-Haq Ali Shah.

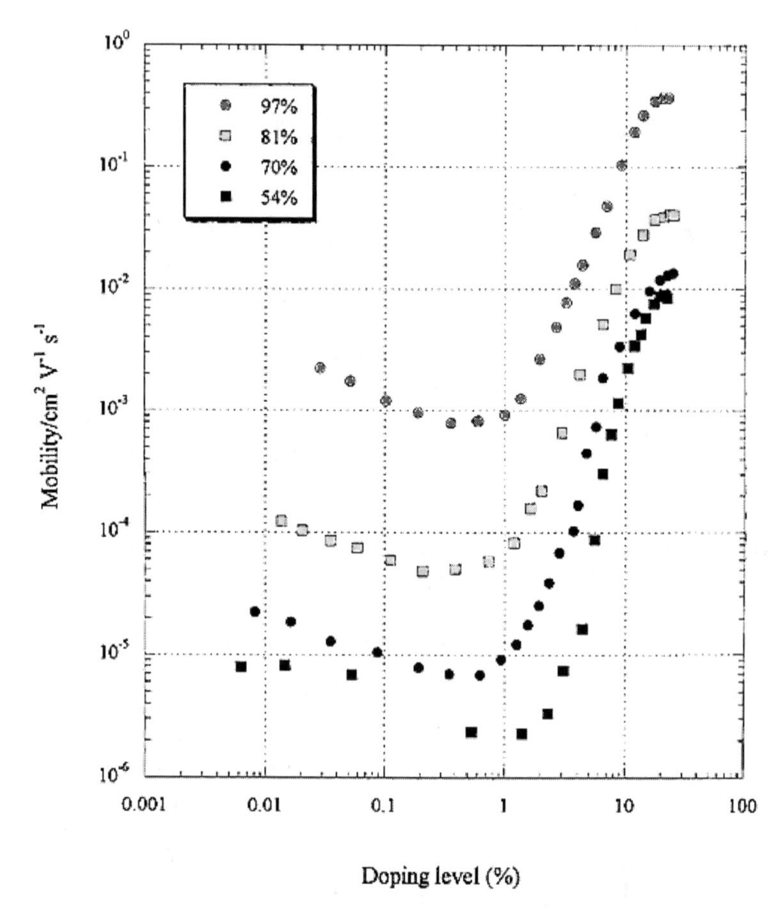

Figure 3.12 Doping-induced change of carrier mobilities in poly(3-hexylthiophene) films with different stacking structures. Reprinted with permission from Ref. 85. Copyright © 2002 Elsevier Science B.V.

Mobilities of positive charge carriers in P3HT films of 54%, 70%, 81%, and 97% in regioregularity were electrochemically measured over a range of doping levels (Fig. 3.12). The results show that better-ordered stacking structures enhance mobilities but affect little features of the mobility change by doping. The common doping level of ca. 1% for the onset of the drastic mobility increase implies that the π–π stacking structures facilitating a charge transport in neutral or lightly doped P3HT films are not responsible for the evolution of metallic conduction in P3HT films.

3.2.2 Chemical Doping: Protonation of Polyaniline

The doping of polyaniline (PANI) is done by protonation (using an acid), while with the other conjugated polymer it is achieved by electron transfer with a dopant or electrochemically (Fig 3.13; Emeraldine base (a), Emeraldine salt (b)).

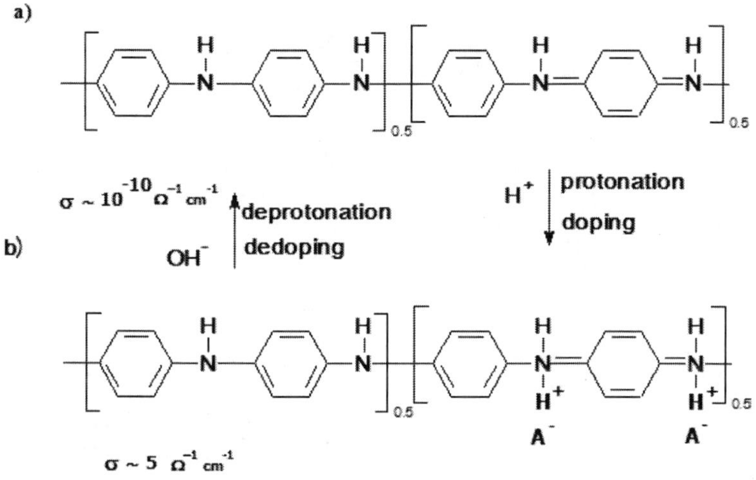

Figure 3.13 Chemical doping and dedoping: protonation and deprotonation of PANI.

CPs can be defined as the cationic and anionic salts of highly conjugated polymers. The cation salts can be obtained by chemical oxidation and electrochemical polymerization. The anion salts of the conjugated polymers are produced by using electrochemical reduction or chemical reduction reagents, i.e., sodium naphthalide. An oxidized CP has electrons removed from the backbone, resulting in a cationic radical. A reduced CP has electrons added to the backbone, resulting in an anionic radical (Fig 3.14).

Figure 3.15 shows the chemical structure of some important conjugated polymers. The conductivities of the organic CPs increase by over 10 orders of magnitude by doping with protonic acids. The doping process is a partial addition or removal of electrons to/from the π system of the polymer backbone.[86]

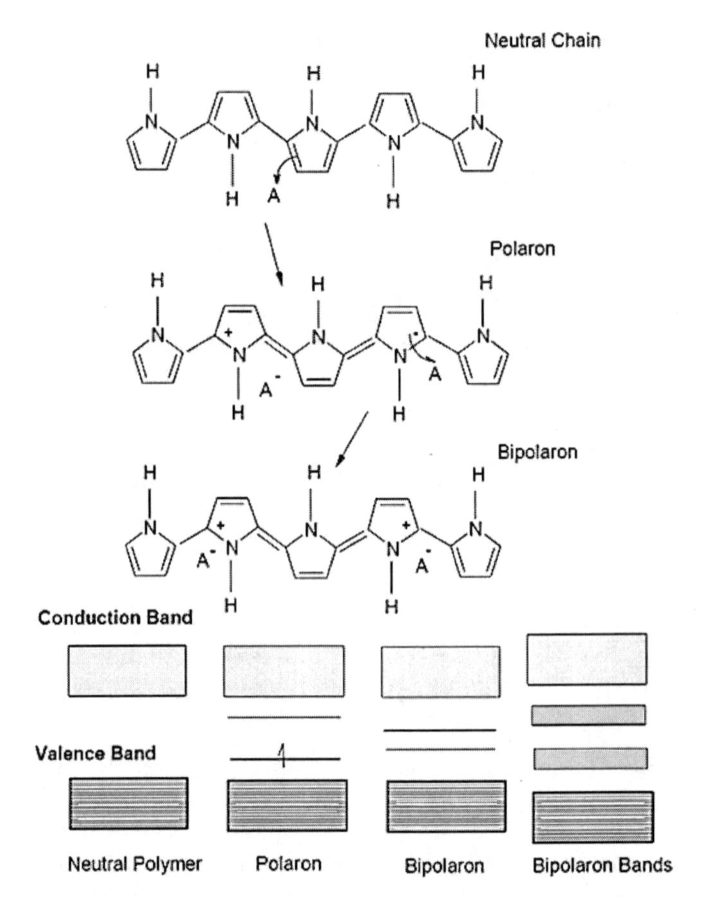

Figure 3.14 Transitions between polaronic and bipolaronic structures in polypyrrole.

The effects of chemical structure on polymer film properties and applications were reviewed.[87] The uses of conductive polymers in the bioanalytical sciences and in biosensor applications were investigated.[88] Synthesis, characterization, and applications of CPs were reported, and the main aspects of CPs in chemical sensors and biosensors were covered.[89,90] The advantages and limitations of conductive polymers for different biomedical applications like tissue engineering, biosensors, drug delivery, and bioactuators were reported. Different preparation methods for conductive polymers and the use of conductive polymers for electromagnetic interference (EMI) shielding applications were reviewed.[91,92]

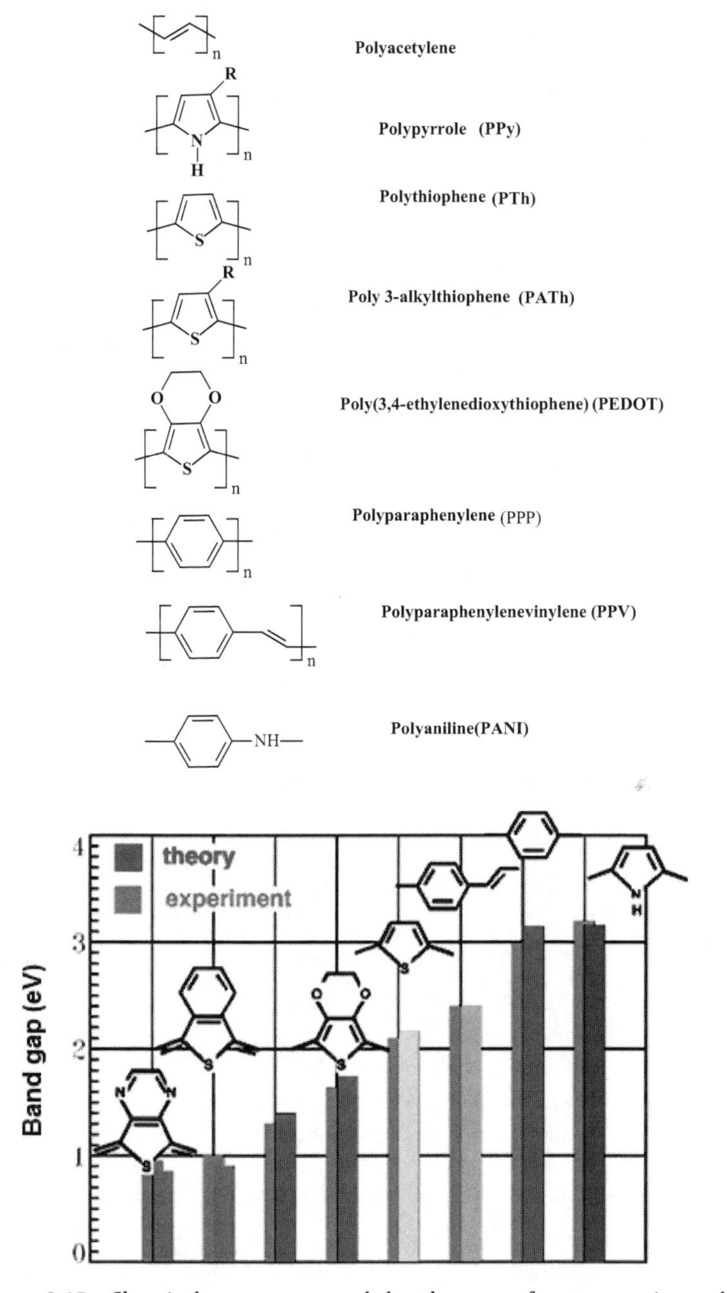

Figure 3.15 Chemical structures and band gaps of some conjugated polymers.

3.3 Applications of Conducting Polymers

CP nanocomposites exhibit multifunctional and unique properties. Therefore, many CP nanocomposites are used in many fields, such as nanoelectronic devices, chemical or biological sensors, catalysis or electrocatalysis, energy, ultracapacitors, ER fluids, and biomedicine.[93,94]

Conjugated polymers were generally unstable in air and it was not easy to prepare films. Proper chemical modifications succeeded to produce stable materials, which can be processed from solution or melt, via a conversion of short-chain precursors.

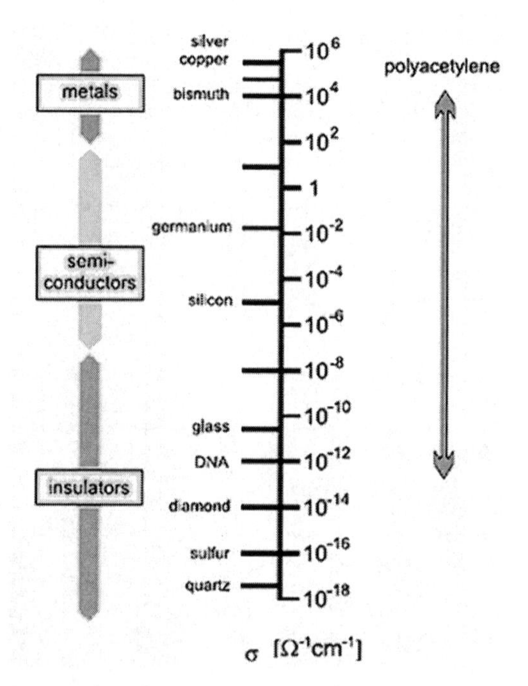

Figure 3.16 Range of conductivities covered by doped polyacetylene (PAC) in comparison with other materials.

To provide a brief introduction to this class of polymer materials, the elementary optical excitations, electroluminescence, and conduction properties (Fig. 3.16) and magnetism of doped samples should be considered.

3.3.1 Electro-Optic Activity

3.3.1.1 Excitons and free charges

In the optical and electrical activity of conjugated polymers several phenomena occur:

- the nature of the excited states that are created when photons are absorbed
- transitions that show up in the subsequent luminescence
- charge carriers that are formed when a current is flowing

The optoelectronic properties of dilute solutions of oligomeric and (broken-conjugation) polymeric PPV chains were studied using optical absorption and (time-resolved) emission spectrophotometry. The following properties were determined: absorption and emission spectra, fluorescence quantum yields and decay times, exciton polarizabilities and dissociation probabilities, charge mobilities, and RC absorption spectra. The experimental results are compared with theoretical calculations of exciton polarizabilities, charge mobilities, and RC absorption spectra.[95]

Electroactive polymers have a very specific chemistry, which may limit the obtention of purely conjugated polymers. Because of limitations on molecular weight and solvents suitable for electrospinning, only a few conjugated polymers such as PANI, poly(dodecylthiophene), and poly[2-methoxy-5-(2'-ethylhexyloxy)-*p*-PPV] have been electrospun. Thus, electrospinning of conjugated polymers is very limited due to the absence of chain entanglement (conjugated backbones are stiffer and offer low or almost no entanglements), which is considered a prerequisite in the electrospinning technique.

Some authors have proposed several modified process approaches to obtain electroactive electrospun nanofibers. One first practical and easy way is to spin a nonconductive polymeric web and after polymerize conductive polymers onto the fiber surface. For example, conductive polyamide-6 (PA-6) nanofibers were prepared by polymerizing pyrrole (Py) molecules directly on the fiber surface of PA-6. First, a solution of PA-6 added with ferric chloride in formic acid was electrospun with average diameter values around 260 nm.

Fibers were then exposed to Py vapor and a compact coating of PPy was formed on the fiber surface, the PPy coating on the fibers turned out to be conductive.

A similar treatment was used to electrospin poly(vinylidene fluoride) (PVDF)/PPy composites, which were prepared by spinning a nonwoven web from a solution of PVDF and $CuCl_2.2H_2O$ in dimethylacetamide (DMAc) and then exposing the spun fibers to Py vapors in order to produce the conductive composites. The electrical conductivity of the PPy composites was affected by the fabrication method and oxidant content in the nonwoven web.

Electrospinning was used for producing a P3HT/[6,6]-phenyl-C61-butyric acid methyl ester (PCBM) electroactive solar cloth (Fig. 3.17). Electrospinning of pure conjugated polymers is limited due to the absence of polymer chain entanglements, that is, for this reason coelectrospinning of poly(3-hexyl thiophene) (P3HT) (a CP) or P3HT/PCBM as the core and poly(*N*-vinyl pyrrolidone) (PVP) as the shell system was proposed. As the solubility of P3HT is limited in chloroform the nozzle was adapted for being clogged by using a coaxial electrospinning setup where the polymer solution is fed through the inner nozzle and pure chloroform is provided through the outer nozzle in order to retard solvent evaporation. This continuous method was suggested to be employed to produce organic-based devices on a massive scale.

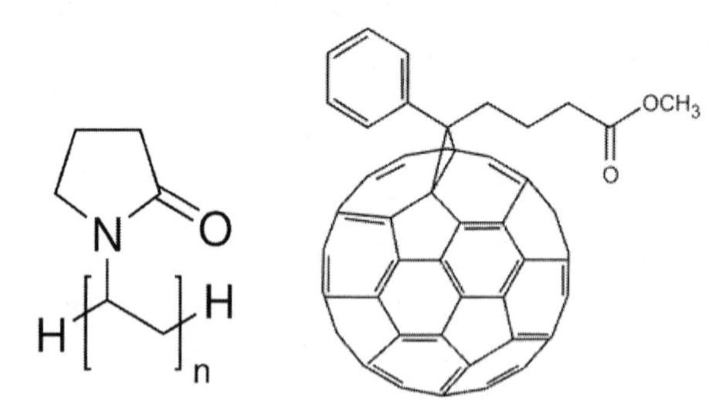

Figure 3.17 Poly(*N*-vinyl pyrrolidone) (PVP) and [6,6]-phenyl-C61-butyric acid methyl ester (PCBM) structures.

Then the PVP shell could be washed, giving rise to conductive P3HT or P3HT/PCBM cloth.

The production of flexible supercapacitors using electroactive fibers obtained by electrospinning, that is, PVP fibers covered with PEDOT by vapor-phase polymerization, was reported.[96] The conductive mats presented elevated electrical conductivity and were separated by a layer of polyacrylonitrile (PAN) in order to assemble an all-flexible capacitor device.

The electrochemical performances of the solid-state supercapacitors are similar to the ones obtained in a liquid electrolyte. Owing to the nanostructure nature of the active materials, an effective wettability by the electrolyte and a limited diffusion length of the doping ions within the polymer structure were observed; however, environmental parameters and fiber diameter reduction can be improved in order to produce more efficient materials.

3.3.2 Tissue Engineering and Sensor Applications

Electrospun conducting fibers are also used in the production of biocompative systems for tissue engineering and biosensors. poly(lactic-co-glycolic acid) (PLGA) electrospun nanofibers coated with PPy were also produced for neural tissue applications.

PPy–PLGA electrospun meshes were used to support the growth and differentiation of rat pheochromocytoma 12 (PC12) cells and hippocampal neurons compared to noncoated PLGA meshes. It was suggested that PPy–PLGA may be suitable as conductive nanofibers for neuronal tissue scaffolds.

Conjugated polymers could be blended with insulating polymers to provide an electrical current to increase cell attachment, proliferation, and migration. Several electrospun PANI and poly (D,L-lactide) (PDLA) mixtures at different weight percentages were studied; only the 75/25 electrospun scaffold exhibited a current of 5 mA with a calculated electrical conductivity of 0.0437 S/cm. Later, primary rat muscle cells were cultured on scaffolds and on tissue culture polystyrene (PS) as a positive control. Although the scaffolds degraded during this process, cells were still able to attach and proliferate on each of the different scaffolds. The cellular proliferation measurements indicated that there is no significant difference between the four groups measured and the conductivity and cellular behavior demonstrate the feasibility of fabricating a biocompatible, biodegradable, and electrically conductive PDLA/PANI scaffold.

HCl-doped poly(aniline-co-3-aminobenzoic acid) (3ABA-PANI) copolymer and poly(lactic acid) (PLA) blend were electrospun in the form of three-dimensional networks with a high degree of connectivity onto glass substrates.

The ability to promote proliferation of COS-1 fibroblast cells over this conductive scaffold was evaluated and these nanofibrous blends are suggested as tissue engineering scaffolds and showed promise as functional wound dressings that may eliminate deficiencies of currently available antimicrobial dressings.

PANI containing gelatin nanofibers for producing conductive scaffolds for tissue engineering purposes was studied; the PANI addition affects the physicochemistry of PANI–gelatin blend fibers since it is biocompatible and supports the cells' (H9c2 rat cardiac myoblast) attachment, proliferation, and growth.

PANI and poly(ε-caprolactone) (PCL) showed similar behavior and their electrospun membranes can be used as platforms to mimic either the morphological or the functional features of the cardiac muscle tissue regeneration. Development of PANI/PCL membranes by electrospinning with controlled texture can create an electrically conductive environment and this environment can stimulate the cell differentiation to cardiomyocites and can be used in the myocardium muscle regeneration.

Composite fibers of poly(*o*-anisidine)–PS was produced by electrospinning for chemical vapor sensing. Sensibility of the composite fibers were tested under water and ethanol vapor, the sensors elements responded better to the high polarity of the solvent. The CSA-doped POA/PS composition seems to be stable under the submitted ambient conditions to ethanol. The sensor could be reused several times without any change in sensing behavior and/or damage to the sensing materials.

A flexible nanotube membrane of PEDOT was produced by electrospinning mediated for ammonia gas detection.

Initially, poly(vinyl alcohol) (PVA) solution was electrospun and further treated with $FeCl_3$ solution to adsorb Fe^{3+} ions on the nanofibers' surface. Later, an EDOT monomer was evaporated and polymerized on the PVA surface, leading to coaxial PVA/PEDOT fibers, which were washed with distillated water, giving 140 nm PEDOT tubes. PEDOT nanotubes achieved electrical conductivity values of 61 S/cm higher than the usual PEDOT nanomaterials produced with $FeCl_3$. PEDOT nanotubes revealed owing faster recovery times than PVA/PEDOT coaxial fibers due to the elevated surface area, demonstrating the possibility of this methodology to produce 1D nanomaterials for sensors applications.

> The electric response of isolated PANI fibers to vapors of aliphatic alcohols depends on the large surface-to-volume ratio, the uniformity of diameter, and the quantity of active material. The sensor constructions are comparable to or faster than those prepared from nanofiber mats of the same polymer. Also, the sensors made from individual fibers exhibit larger responses, especially for bigger alcohol molecules, and also show true saturation upon exposure and removal of the alcohol vapor. The response of sensors made from electrospun nanofibers to small alcohol molecules is opposite to that observed for cast nanofiber mats, which can be related to the doping process used in the preparation of the polymer in either case.*

3.4 Alkyl-Substituted Polythiophenes

Thiophene (Th) and its derivatives are polymerized by chemical and electrochemical methods, especially poly(3-methylthiophene) (3-mTh), which was investigated because of its remarkable solid-state properties, including thermochromism, electrochromism, luminescence, and photoconductivity. Removal of two electrons (*p*-doping) from a polyalkylthiophene (PAT) chain produces a bipolaron structure (Fig. 3.18).

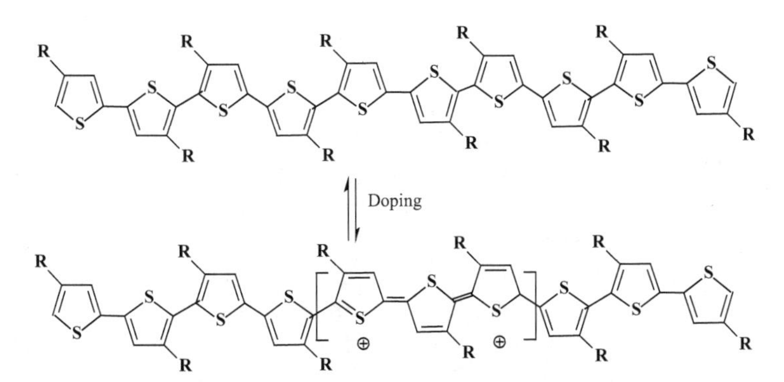

Figure 3.18 Removal of two electrons (*p*-doping) from a polyalkylthiophene chain produces a bipolaron.

*Excerpt reprinted with permission from Ref. 97, Copyright 2007, Elsevier.

3.4.1 Regioregularity

The asymmetry of 3-substituted thiophenes results in three possible couplings when two monomers are linked between the 2- and the 5-positions. These couplings are:

- 2,2′, or head–head (HH)
- 2,5′, or head–tail (HT)
- 5,5′, or tail–tail (TT)

These three diads can be combined into four distinct triads, as shown in Fig. 3.19.

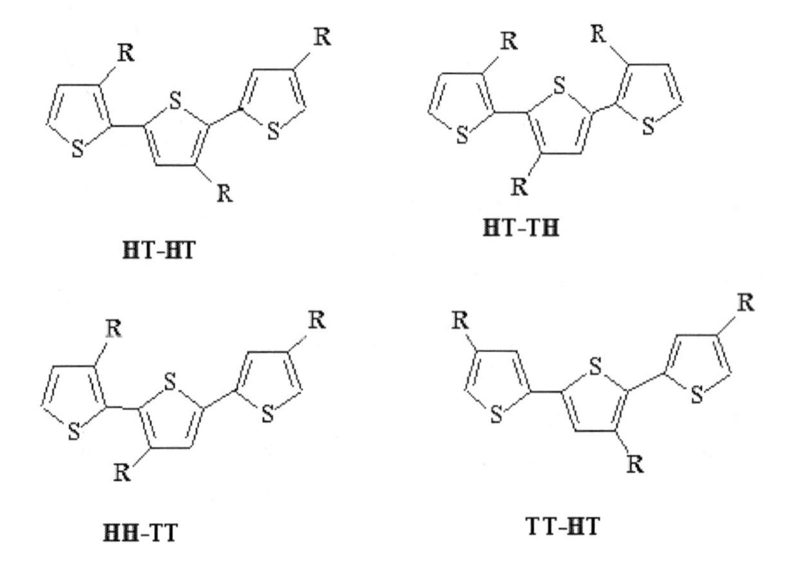

Figure 3.19 The four possible triads resulting from coupling of 3-substituted thiophenes.

The triads are distinguishable by nuclear magnetic resonance (NMR) spectroscopy, and the degree of regioregularity can be estimated by integration.

A regiorandom copolymer of 3-methylthiophene and 3-butylthiophene possessed a conductivity of 50 S/cm, while a more regioregular copolymer with a 2:1 ratio of HT to HH couplings had a higher conductivity of 140 S/cm.[98] Films of regioregular poly[3-(4-octylphenyl)thiophene] (POPT) with greater than 94% HT content possessed conductivities of 4 S/cm compared to 0.4 S/cm for regioirregular POPT.

Comparison of the thermochromic properties of regioregular polymers showed strong thermochromic effects, and the absorbance spectra of the regioirregular polymers did not change significantly at elevated temperatures. This was likely due to the formation of only weak and localized conformational defects.The fluorescence absorption and emission maxima of P3HTs occur at increasingly lower wavelengths (higher energy) with increasing HH dyad content. The difference between absorption and emission maxima, the Stokes shift, also increases with HH dyad content, which was attributed to greater relief from conformational strain in the first excited state.

3.4.2 Solubility

Unsubstituted polythiophenes (PThs) are conductive after doping and have good stability compared to some other conjugated polymers, such as polyacetylene, but are intractable and soluble only in solutions like mixtures of arsenic trifluoride and arsenic pentafluoride. Organic-soluble PThs were obtained by using a nickel-catalyzed Grignard cross-coupling and synthesized two soluble PThs, poly(3-butylthiophene) and poly(3 methylthiophene-co-3-octylthiophene), which could be cast into films and doped with iodine to reach conductivities of 4–6 S/cm. Poly(3-butylthiophene) and P3HT were synthesized electrochemically and chemically and cast into films. The soluble PATs demonstrated both thermochromism and solvatochromism in chloroform and 2,5-dimethyltetrahydrofuran.[99]

Also the synthesis of water-soluble sodium poly(thiophene-alkane-sulfonate) are reported. In addition to conferring water solubility, the pendant sulfonate groups act as counterions, producing self-doped CPs. Substituted PThs with tethered carboxylic acids, acetic acids, amino acids, and urethanes are also water soluble.[99]

More recently, poly(3-[perfluorooctyl]thiophene)s soluble in supercritical carbon dioxide were electrochemically and chemically synthesized. Unsubstituted oligothiophenes capped at both ends with thermally-labile alkyl esters were cast as films from solution and then heated to remove the solubilizing end groups. Atomic force microscopy (AFM) images showed a significant increase in long-range order after heating.[99]

The oxidation of a monomer produces an RC, which can then couple with a second RC to form a dication dimer or with another monomer to produce an RC dimer. Deposition of long, well-ordered chains onto the electrode surface is followed by growth of either long, flexible chains or shorter, more crosslinked chains, depending upon the polymerization conditions. The quality of an electrochemically prepared PTh film is affected by a number of factors. These include the electrode material, current density, temperature, solvent, electrolyte, presence of water, and monomer concentration. Two other important but interrelated factors are the structure of the monomer and the applied potential. The potential required to oxidize the monomer depends upon the electron density in the thiophene ring π system. Electron-donating groups lower the oxidation potential, while electron-withdrawing groups increase the oxidation potential. Thus, 3-methylthiophene polymerizes in acetonitrile and tetrabutylammonium tetrafluoroborate at a potential of about 1.5 V versus a saturated calomel electrode (SCE), while unsubstituted thiophene polymerizes at about 1.7 V versus an SCE. Steric hindrance resulting from branching at the α-carbon of a 3-substituted thiophene inhibits polymerization. This observation leads to the so-called *polythiophene paradox*: the oxidation potential of many thiophene monomers is higher than the oxidation potential of the resulting polymer.[99]

The polymer can be irreversibly oxidized and decomposed at a rate comparable to the polymerization of the corresponding monomer, which is one of the disadvantages of electrochemical polymerization and limits its application for many thiophene monomers with functional side groups.

3.4.3 Chemical Oxidative Polymerization

Studies have been conducted to improve the yield and quality of the product obtained using the oxidative polymerization technique. In addition to ferric chloride, other oxidizing agents, including ferric chloride hydrate, ceric ammonium nitrate and ceric ammonium sulfate, copper perchlorate, and iron perchlorate, have also been used to polymerize 2,2′-bithiophene. Controlled addition of ferric chloride to the monomer solution produced POPTs with ~94% HT content. Precipitation of $FeCl_3$ in situ treatment produced significantly higher yields and monomer conversions than adding the monomer directly to the crystalline nickel catalyst. Higher molecular weights were reported when dry air was bubbled through the reaction mixture during polymerization. Soxhlet extraction after polymerization with polar solvents was found to effectively fractionate the polymer and remove the residual catalyst. Using a lower ratio of catalyst to monomer (2:1 rather than 4:1) may increase the regioregularity of poly(3-dodecylthiophene)s. Higher yields of soluble poly(dialkylterthiophene)s in carbon tetrachloride rather than chloroform were found, which were attributed to the stability of the radical species in carbon tetrachloride. A higher-quality catalyst, added at a slower rate and at reduced temperature, was shown to produce high-molecular-weight PATs with no insoluble polymer residue, and increasing the ratio of catalyst to monomer increased the yield of poly(3-octylthiophene), and a longer polymerization time also can increase the yield.

The possible mechanisms of the oxidative polymerization using ferric chloride has been proposed, as shown in Fig. 3.20.

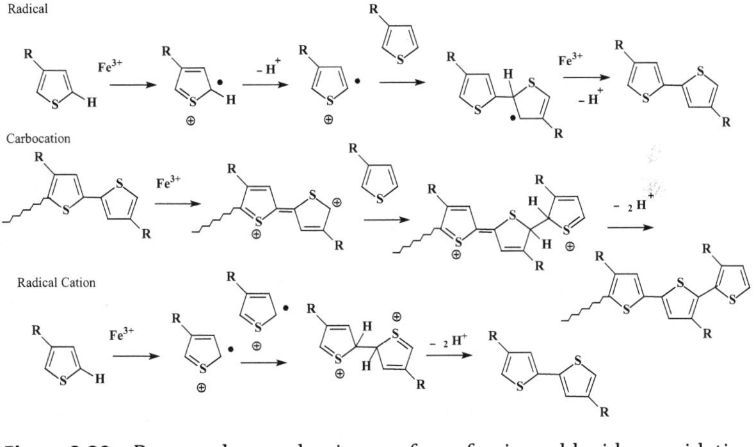

Figure 3.20 Proposed mechanisms for ferric chloride oxidative polymerizations of thiophenes.

The mechanisms are based on two assumptions. First, since polymerization was observed only in solvents where the catalyst was either partially or completely insoluble (chloroform, toluene, carbon tetrachloride, pentane, and hexane, not diethyl ether, xylene, acetone, or formic acid), it was concluded that the active sites of the polymerization must be at the surface of solid ferric chloride. Therefore, there were possibilities of either two RCs reacting with each other or two radicals reacting with each other because the chloride ions at the surface of the crystal would prevent the RCs or radicals from assuming positions suitable for dimerization. Second, using 3-methylthiophene as a prototypical monomer, the quantum mechanical calculations determine the energies and the total atomic charges on the carbon atoms of the four possible polymerization species (neutral 3-methylthiophene, the RC, the radical on carbon 2, and the radical on carbon 5).[99]

Since the most negative carbon of the neutral 3-methylthiophene is also carbon 2, and the carbon with the highest odd electron population of the RC is carbon 2, it was concluded that an RC mechanism would lead to mostly 2–2, HH links. The calculated total energies of the species with the radicals at the 2 and 5 carbons indicated that the latter was more stable by 1.5 kJ/mol. Therefore, the more stable radical could react with the neutral species, forming head-to-tail couplings, as shown in Fig. 3.21.[99]

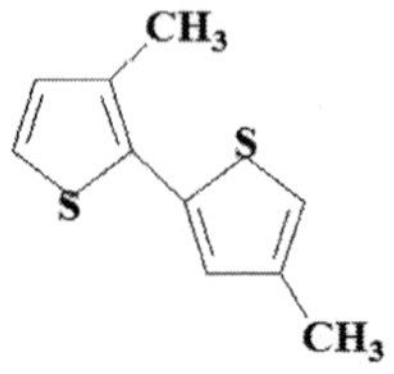

Figure 3.21 Head-to-tail couplings.

An alternative mechanism of the polymerization of POPT with ferric chloride with a high degree of regioregularity was found when the catalyst was added to the monomer mixture slowly. It was proposed that, given the selectivity of the couplings and the strong oxidizing conditions, the reaction could proceed via an RC mechanism. The radical mechanism was reported for thiophene, which could be polymerized by ferric chloride in acetonitrile as the solvent in which the catalyst is completely soluble. The kinetics of thiophene polymerization also seemed to contradict the predictions of the radical polymerization mechanism but an RC mechanism analogous to electrochemical polymerization was also a possibility.

Due to the difficulties of studying a system with a heterogeneous, strongly oxidizing catalyst that produces difficult-to-characterize rigid-rod polymers, the mechanism of oxidative polymerization is not easy to decide. However, the RC mechanism seems to be the likely route for PTh synthesis.

Electropolymerization of monomers requires high oxidation potentials, and during the past decade, boron trifluoride diethyl etherate (BFEE) has been used as a solvent and a supporting electrolyte.[100] Since BFEE is a strong Lewis acid its acidity should be lowered by dissolving it in a coordinating solvent with an acetonitrile or diethyl ether. Films obtained by the BFEE-coordinating solvent system were flexible and free standing compared to CPs obtained in common organic solvents, having a fragile structure and a limited processability.

3-mTh was electrodeposited in the form of both a homopolymer and a copolymer to examine electrochemical, spectroelectrochemical, and electrochromic properties in common electrochemical solvents. A 3-mTh-activated carbon hybrid capacitor with a gel polymer electrolyte exhibited a high capacitance value of 18.54 F/g.

3.4.4 Catalyst for the Deprotonation of Oligomers of 3-Methylthiophene

The interaction between 3-mTh and a Lewis acid solvent BFEE markedly lowered the electrochemical oxidation potential of the monomer where the formation of RCs and dehydrogenation eases by the interaction of BFEE. The electrophilic property of the Lewis acid BFEE catalyzes the deprotonation of oligomers of 3-methylthiophene on the electrode, resulting in a very low oxidation potential. The lower oxidation potential of 3-methylthiophene in BFEE may reduce the content of α-β'linkages between monomer units, thereby imparting more linearity to the overall structure of polymethylthiophene compared to PTh, resulting in improved stereoregularity.

As a Lewis acid, BFEE can interact with the aromatic ring, which can lower the oxidation potential during the electrochemical polymerization. Due to acidity of BFEE, additive polymerization can occur and nonconjugated polymers can result.

Electrochemical polymerization of 3-methylthiophene (Fig. 3.22) follows an RC formation and the growth of 3-mTh is accomplished by electrophilic attack of an RC at the end of the growing chain on the neutral monomer unit, which is followed by oxidation and deprotonation steps. The process is suggested as neither a classical step growth nor a classical chain polymerization but something in between.[100]

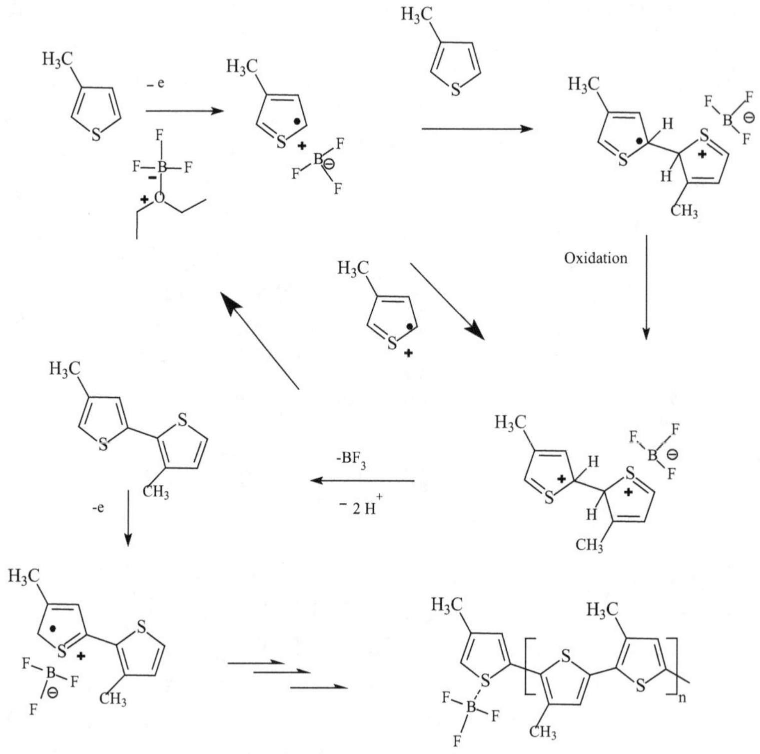

Figure 3.22 Electrochemical polymerization of 3-methylthiophene.

3.5 Poly(3,4-Ethylenedioxythiophene) Composites

PEDOT is an important π-conjugated CP, which is currently being investigated for use in many fields, such as antistatic and anticorrosion materials, artificial muscles, electrode materials in batteries, supercapacitors, display devices, and biosensors. Although various PEDOT nanomaterials, such as nanofibers, nanospheres, nanotubes, and nanorods, have been prepared and studied, PEDOT films are studied as catalyst supports for Pt or Pd nanoparticles for electro-oxidation of either methanol or ethanol,[101] which can be potentially used in direct methanol fuel cells (DMFCs) or sensors to some chemicals, such as nitrite, bromate, oxygen, and hydrogen peroxide. As many applications of the PEDOT derivatives are related to its microstructure and electrochemical activity, studies on the film surface control and physico-chemical properties are very important.

The electrospinning method was applied to produce nanofibers of PEDOT-PSS/poly(vinyl acetate) (PVAc), which can be used for adhesives and coatings. An aqueous emulsion of PEDOT-PSS was added to the PVAc/*N,N*-dimethylformamide (DMF) solution. Composites that include different amounts of PEDOT-PSS were prepared.

3.5.1 Poly(3,4-Ethylenedioxythiophene)/Polystyrene Sulfonic Acid

The presence of PSS in the PEDOT/PSS yields a water-soluble polyelectrolyte with a good film forming properties, high conductivity (ca. 10 S/cm), high light transmissivity, and good stability. Films of PEDOT-PSS can be heated in air at 100°C for over 1000 hours with only a small change in conductivity. PEDOT/PSS was used as an antistatic coating in photographic films, but recently new applications have been implemented, such as electrode material in capacitors and material for through-hole plating of printed circuit boards. PEDOT and PSS contain sulfur group as PEDOT in thiophene ring, but in PSS it is in sulfonate moiety.

Chemical polymerization of EDOT derivatives can be carried out using several methods. In the classical method $FeCl_3$ (iron (III) chloride) or $Fe(OTs)_3$ (iron(III) *p*-toluenesulfonate) are used as an oxidizing agent. The most useful polymerization method for EDOT utilizes the polymerization of EDOT in an aqueous polyelectrolyte, i.e., PSS solution using $Na_2S_2O_8$ (sodium persulfate) as the oxidizing agent. Carrying out this reaction at room temperature, aqueous PEDOT/PSS results in a dark-blue solution.

3.5.2 Conductivity Enhancement of PEDOT-PSS

The conductivity of a PEDOT-PSS film can be enhanced by more than 2 orders of magnitude by the addition of polar group containing compounds like ethylene glycol, into an aqueous solution of PEDOT-PSS. The additive induces a conformational change in the PEDOT chains of the PEDOT-PSS film.

Both coil and linear or expanded-coil conformations exist in untreated PEDOT-PSS films, whereas the linear or expanded-coil conformation becomes dominant in high-conductivity PEDOT-PSS films. This conformational change results in an increase in the intrachain and interchain charge carrier mobility, so the conductivity is enhanced.[102]

3.5.3 Charge Transport in the Conducting Polymer PEDOT-PSS

CPs have rarely been produced in a form ordered enough to exhibit a small positive temperature coefficient of resistivity. This so-called metallic behavior is normally not seen in the transport properties of as-grown CPs, like PPys, PThs, or PANIs, where the negative temperature coefficient of the resistivity is generally attributed to hopping.

The charge transport mechanism in the case of PEDOT-PSS can be explained by the concept of the charging-energy-limited tunneling model, proposed for highly disordered CPs. In this model, conduction is proposed as from tunneling between small conducting grains separated by insulating barriers. This model is an extension model of granular metals which focuses on the disorder present in the polymer and the polaronic ground state characteristic in most CPs. This model considers that the conducting clusters are highly doped "polaronic islands" generated by heterogeneities in the doping distribution. The dopant centers act as bridges between neighboring chains by improving the charge carrier transport.

3.5.4 Morphology of PEDOT-PSS

The structure and morphology of thin films might be different than those of bulk material. For PEDOT, thin spin-cast and bulk solution-cast films show different kinds of orientation, although they have the same basic crystalline structure. Grazing-incidence X-ray diffraction studies have indicated that the dopant ions form distinct planes, which alternate with stacks or lamellae of polymer chains. The material is very anisotropic, with the planes of the dopants and of the stacks of polymer chains parallel to the substrate. It is in a paracrystalline state, with a small size of the individual paracrystalline regions. This model corroborates well with the strong optical anisotropy observation. Despite the observed anisotropy in the optical response, possible anisotropies in the conductivity were systematically addressed.[103]

3.6 Copolymers of PEDOT

- Copolymers of PEDOT have the electronic and optical properties of PEDOT but can be processed from nonacidic organic solutions. Some commercial materials (TDAE) have good mechanical properties, low acidity, and wet organic substrates without the use of binders or additives.
- The new materials are block copolymers of PEDOT and a flexible polymer such as polyethers, polysiloxanes, polyesters, or polyacrylates.
- The blocks of PEDOT and the flexible polymer are covalently bonded in a variety of ways that lead to the formation of different geometries among the blocks (Fig. 3.23).[104]
- Block copolymers of PEDOT with poly(ethylene glycol) (PEG), poly(propylene glycol) (PPG), polydimethylsiloxane (PDMS), and acrylic derivatives are available (Fig. 3.23).

PEDOT, which has reactive methacrylate end groups, is dispersible in an organic solvent such as nitromethane or propylene carbonate. Films cast from these solutions are smooth and scratch resistant and have good adhesion properties on many substrates. And thin films appear transparent blue with an optical transmittance up to 75%.[104]

Depending on what anion is used to dope these copolymers, bulk conductivity is measured in the 0.1–1.0 S/cm range (Aedotron™-C) or 10–60 S/cm (Aedotron™-C3) (Fig. 3.24; Table 3.3).[104]

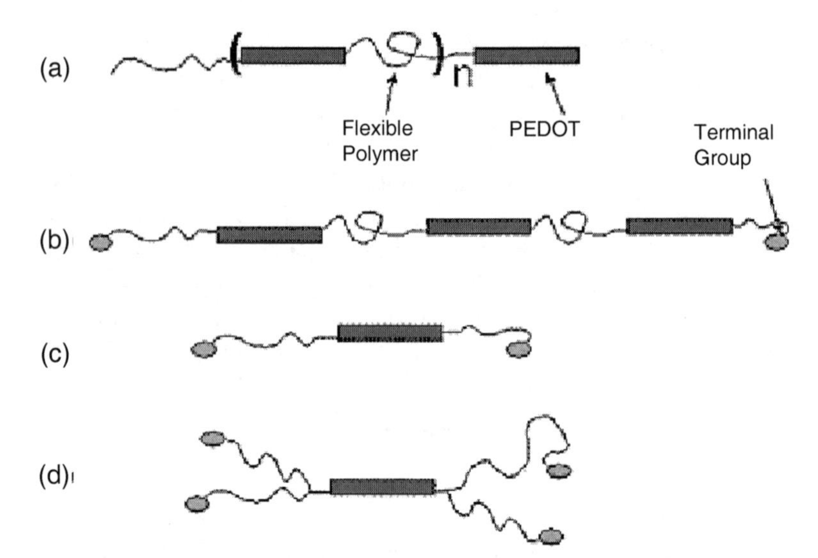

Figure 3.23 Block copolymers of PEDOT with flexible polymers.

Figure 3.24 Aedotron^{TM} C3-NM: Aldrich.

Table 3.3 Comparison of two commercially available Aedotron™ conducting copolymers

Material	Particle Size in Suspension (nm)	Bulk Conductivity (S/cm)	Sheet Resistance (Ω/square)	Average Transmittance (%T)	RMS Roughness of Spin-Cast Thin Films (nm)
Aedotron™ C-NM 649805	600–1000	0.1–2	104–105	70%–85%	40
Aedotron™ C3-NM 687316	200–600	10–60	600–3000	70%–85%	10

Cast films can exhibit a wide range of sheet resistance, depending on film thickness, preparation method, and material choice.

The graph in Fig. 3.25 shows a plot of the visible transmittance of cast thin films of Aedotron^{TM}-C that were dip-coated on polycarbonate from a propylene carbonate dispersion and dried at 80°C. The sheet resistance of each film is also shown in the graph.

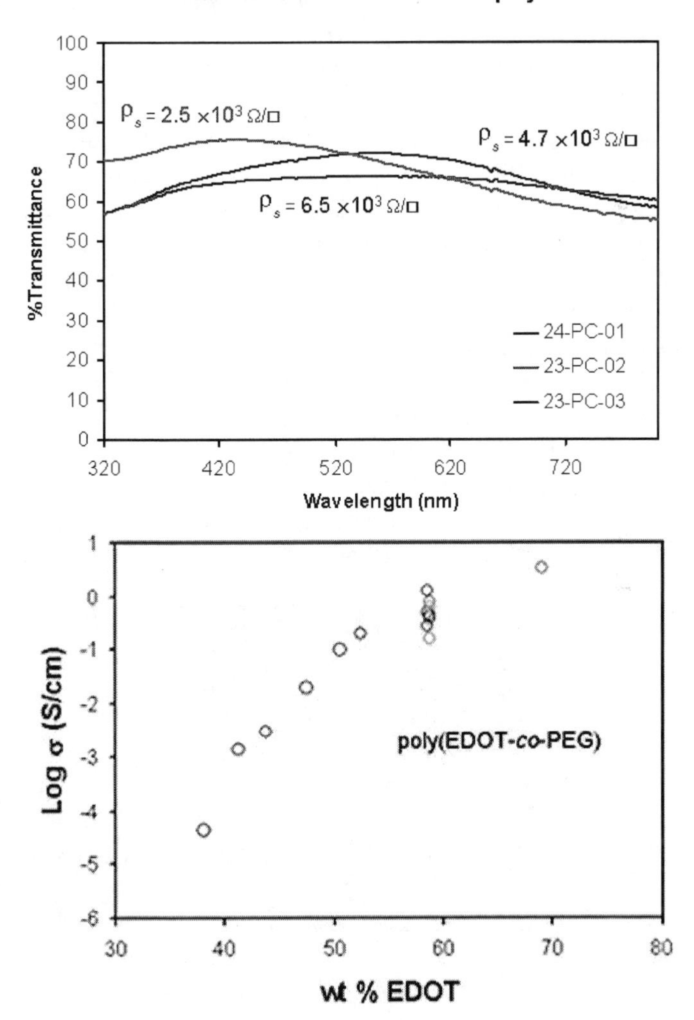

Figure 3.25 Transmittance and conductivity of poly(EDOT-co-PEG) [104].

3.7 Supercapacitors

A supercapacitor is a promising energy storage device that is positioned between a conventional electrolytic capacitor and a rechargeable battery. Moreover, high power, high energy, and long-term reliability of a supercapacitor enable this component to be used in various applications as a backup power unit, an auxiliary power unit, instantaneous power compensation, peak power compensation, and energy storage as well (Fig. 3.26).

Supercapacitors are electrochemical double-layer capacitors identified by high energy and power density. Additionally, they contain two electrodes surrounded by an electrolyte and separated with a separator. Since the loading capacity depends on the electrode surface, significant performance enhancement of supercapacitors will be achieved through nanostructuring and the associated surface extension. From the nanotechnological point of view, new materials as nanoporous substances are perfectly suitable as graphitic electrode materials in supercapacitors due to their extremely high inner surface, adjustable pore distribution, and pore diameters.[105,106]

Coupling the ultrasmall separation distance with a relatively large surface area, in ultracapacitors the ratio of available surface area to charge separation distance h makes capacitors "ultra." The ability to hold opposite electrical charges in static equilibrium across molecular spacing is the important feature.

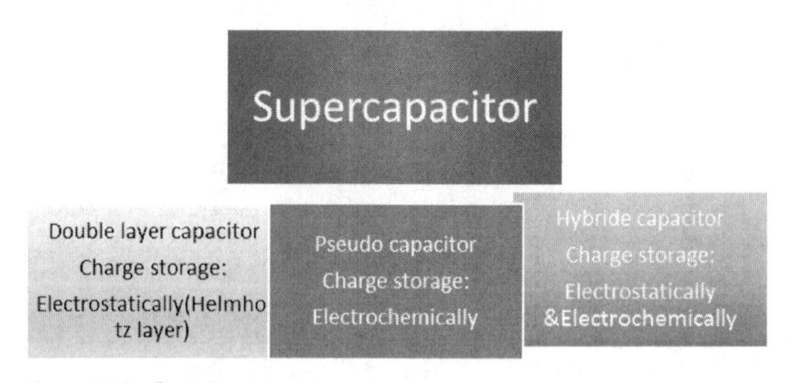

Figure 3.26 Capacitor types.

Three main factors determine how much electrical energy a capacitor can store: the electrode surface area, the electrode separation distance, and the properties of the insulating layer separating the electrodes (Fig. 3.27).

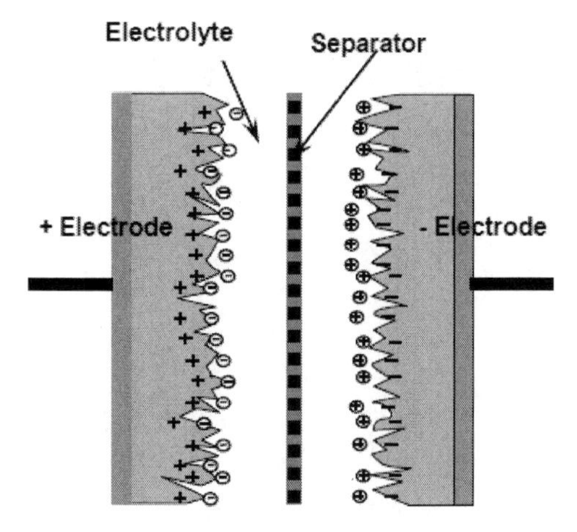

Figure 3.27 Ultracapacitor charge separation.

Different types of electrode materials have been used for the development of redox supercapacitors. Especially PTh derivatives have received considerable attention due to their applicability in both aqueous and organic electrolyte solutions.

PEDOT derivatives show good conductivity and high stability in the oxidized state[107] and they are considered to be useful for electrochemical supercapacitor applications.[108,109] In the polymerization of EDOT, the electron-donating oxygen atoms in the 3- and 4-position not only reduce the oxidation potential of the aromatic ring but also prevent α-β coupling during the chain growth. The rigidity of the ethylene bridge minimizes steric distortion effects, leading to polymers of high regularity with well-ordered nanostructures.

The tetrahedral substitution pattern of quaternary blocked poly(3,4-propylenedioxythiophene) (PProDOT) derivatives[110,111] is directing the alkyl substituents above and below the plane of the π-conjugated chain, allowing high doping levels and inhibiting π stacking of the central carbon of the propylene bridge (Fig. 3.28).

The value of double-layer capacitance C_{dl} can be determined by electrochemical impedance spectroscopy (EIS), which is a key factor to characterize electrochemical systems as components of double-layer capacitors. The electrochemical response of a conductive or semiconductive interface provides characteristic information related to electronic, chemical, and mass transfer processes in the frequency domain.

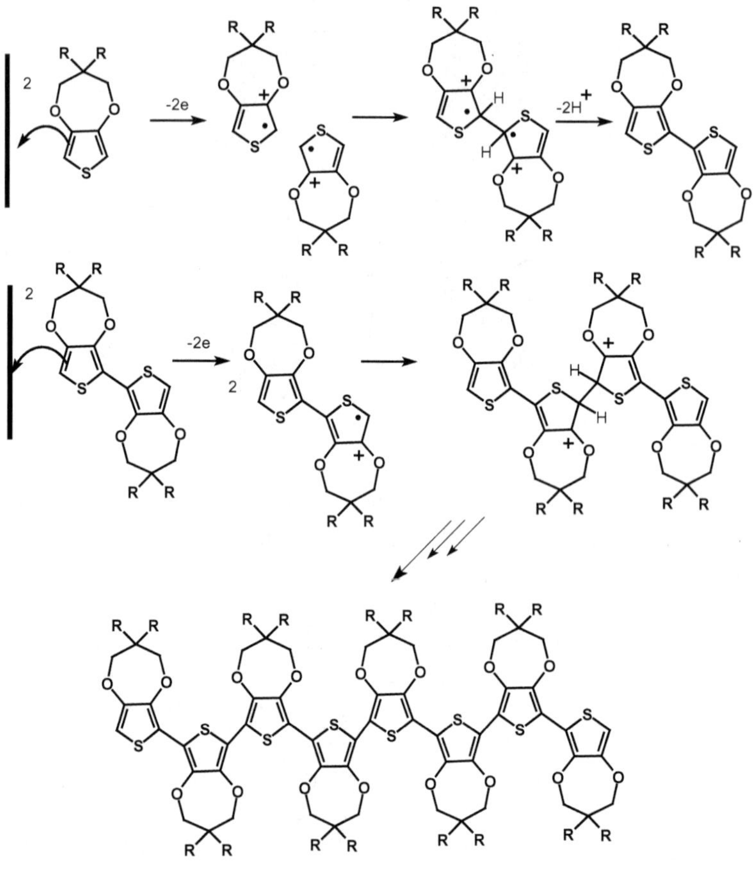

Figure 3.28 Electrochemical polymerization of ProDOT derivatives.

3.8 Electrochemical Impedance Spectroscopy

EIS is an efficient electrochemical technique for studying a variety of chemical, electrochemical, and surface reactions. This technique is used due to the ability of the method to give information on both the bulk and the interfacial properties of polymer-coated electrodes. EIS is the technique which is measuring impedance (complex resistance) response over a wide range of frequencies and providing so much information due to the response of the system changes with frequency. Thus, a broad range of information can be obtained in a single measurement. The impedance is described by an equation analogous to Ohm's law ($R = E/I$), where R is the resistance (ohms), E is the direct current (DC) potential (V) and I is the current (amperes) in DC. This relationship is only valid for an ideal resistor, which does not exist in reality. Impedance is a measure of the ability of a circuit element to resist electrical current flow. Impedance measurements are made by applying a small alternating current (AC) potential of known frequency (ω) with a small amplitude (E_0) to a system and measuring the current and phase difference φ of the concomitant electrical current that develops across it. φ is the phase difference between the sinusoidal voltage and sinusoidal current. $\varphi = 0$ for purely resistive behavior. The consideration of phase shift makes the impedance a more general concept than resistance. Since the impedance values are complex numbers, they can be represented in two different manners, that is, Nyquist and Bode plots. The Nyquist plot shows the data as real (Z') versus imaginary (Z''), while the Bode plot shows the amplitude and phase values over a range of frequencies. In the Nyquist plot, each point gives the characteristics of the complex impedance at the given frequency.

Chapter 4

Polymerization Techniques

4.1 Polymerization Techniques

Conjugated polymers demonstrate reasonable electrical conductivity which can be used in antistatic coatings, electronic devices, batteries, electromagnetic interference (EMI) shielding, electrochromic devices, optical switching devices, sensors, and textiles.[112]

Conjugated polymers, such as polypyrrole, polythiophene, polyaniline (PANI) and its derivatives, are the most promising class of organic-semiconducting materials due to their better electrochemical behavior, electrochromism, ease of doping, and synthesis. But, on the other hand the low processability due to its low solubility is a problem.

Aniline was the first example of the conjugated polymers doped by proton, it can be chemically oxidized by ammonium peroxydisulfate (APS). PANI has the advantages of easy synthesis, low-cost, proton doping mechanism, and is controlled by both oxidation and protonation state. Polyanilines are commonly prepared by the chemical or electrochemical oxidative polymerization of the respective monomers in acidic solution. But, other polymerization techniques have also been developed, including:

Nanofibers of Conjugated Polymers
A. Sezai Sarac

ISBN 978-981-4613-51-4 (Hardcover), 978-981-4613-52-1 (eBook)
www.panstanford.com

1. Chemical polymerization
2. Electrochemical polymerization
3. Photochemically initiated polymerization
4. Enzyme-catalyzed polymerization
5. Polymerization using electron acceptors

Comparison to other film preparation methods, the solution preparation for the electrospinning method through chemical polymerization seems to be more convenient and efficient method.

4.2 Synthesis and Characterization of PEDOT and PEDOT/PSS

The synthesis of poly(3,4-ethylenedioxythiophene (PEDOT) derivatives can be classified into three different types of polymerization reactions:[113]

- Oxidative chemical polymerization of 3,4-ethylenedioxythiophene (EDOT)-based monomers
- Electrochemical polymerization of EDOT-based monomers
- Transition metal–mediated coupling of dihalo derivatives of EDOT

4.2.1 Oxidative Chemical Polymerization of EDOT-Based Monomers

Chemical polymerization of EDOT and its derivatives can be carried out using several methods and oxidants. The classical method employs oxidizing agents such as $FeCl_3$ or $Fe(OTs)_3$.

Chemical polymerization of alkylated or alkoxylated EDOT derivatives[114,115] results in regiorandom PEDOT derivatives soluble in common organic solvents such as chloroform and tetrahydrofuran (THF). These polymers can be characterized by standard methods; for example, gel permeation chromatography (GPC) determined M_w values for a PEDOT range from 10,000–25,000 g/mol.

The second polymerization method of EDOT is also possible by utilizing $FeIII(Ots)_3$, at elevated temperature, in combination with imidazole as a base, resulting in a black insoluble and infusible

PEDOT film that after rinsing with water and *n*-butanol exhibits conductivities of up to 550 S/cm.[116]

Another polymerization method for EDOT is the so-called BAYTRON P synthesis, which was developed at Bayer AG[117] and modified by ELECON, Inc. This method utilizes the polymerization of EDOT in an aqueous polyelectrolyte (most common polystyrene sulfonic acid [PSS]) solution using $Na_2S_2O_8$ as the oxidizing agent.

The reaction at room temperature results in a dark-blue, aqueous PEDOT/PSS dispersion, which is commercially available from Bayer AG under the trade name BAYTRON P and from ELECON under the trade name ELEFLEX13 2000. After drying of BAYTRON P and ELEFLEX 2000, the remaining PEDOT/PSS film is highly conducting, transparent, mechanically durable, and insoluble in any common solvent.

The structural model for tosylate-doped PEDOT presented a contribution to the structural characterization of PEDOT by using de Leeuw's chemical method extended to a surface-confined polymerization. Thin PEDOT films were formed that were subsequently studied with grazing incidence X-ray diffraction using synchrotron radiation. From these studies it was concluded that the material is very anisotropic and these thin films exhibit a limited crystalline order.

PEDOT's electronic structure by X-ray and ultraviolet photoelectron spectroscopy was also studied as well as by spectroscopic ellipsometry. These results suggest that PEDOT prepared in this manner can be seen as an anisotropic metal.

Fourier transform infrared spectrophotometry (FTIR) spectra of the neutral polymer PEDOT indicates a regular structure formed via α-α coupling of thiophene rings.[118] The polymer shows two absorption bands at 341 nm and 413–419 nm in an *N*-methyl-2-pyrrolidone (NMP) solution in the ultraviolet-visible (UV-Vis) luminescence with peak at ca. 552 nm.

4.3 PEDOT as an Electrode Material for Solid Electrolyte Capacitors

A second application for PEDOT derivatives is their use as electrodes in solid electrolyte capacitors, as one example illustrated schematically in Fig. 4.1. A tantalum solid electrolytic capacitor (Ta/MnO_2, capacitor), using manganese dioxide (MnO_2) as a counter electrode, has been employed because of its high capacitance per unit volume. However, the capacitor has inferior frequency characteristics due to the relatively low conductivity of MnO_2, some tantalum solid electrolytic capacitors employing polypyrrole and polyaniline as the counter electrodes were proposed. Recently, poly(3, 4-ethylenedioxythiophene) (PEDOT) has been developed as another new type of tantalum solid electrolytic capacitor (Fig. 4.1).[119]

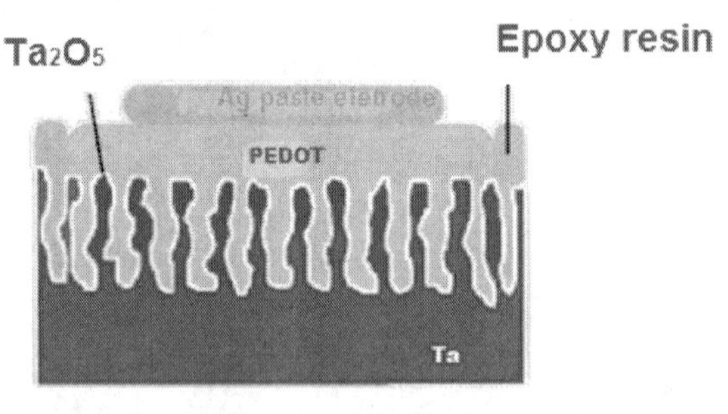

Figure 4.1 Setup of a PEDOT-coated Ta/Ta_2O_5 capacitor.

To prepare these capacitors, the surface of an aluminum or tantalum electrode was roughened by etching or metal powder can be sintered to obtain a high-surface-area electrode with high capacitance. A thin film of metal oxide, which would serve as a dielectric, can be applied to the electrode by anodization. A counter electrode was ultimately applied to complete the capacitor. In the preparation of conventional capacitors, this last step was performed by multiple impregnations with manganese nitrate solution and pyrolysis was applied to form electrically conductive manganese dioxide.

4.3.1 Preparation of Poly(Vinyl Acetate)/Poly(3,4-Ethylenedioxythiophene) Poly(Styrene Sulfonate) Composites

Poly(vinyl acetate) (PVAc) 0.1 g was dissolved in 10 mL *N,N*-dimethylformamide (DMF). PVAc solutions of PEDOT and PEDOT/PSS were prepared and characterized to indicate the differences between PEDOT and PEDOT-PSS/PVAc composites.[120] The series of polymer solutions were prepared with different amounts of 1.3 wt.% PEDOT-PSS aqueous solution, which were 0.25 g, 0.50 g, 0.75 g, 1.00 g, 1.25 g, and 1.5 g each in 10 mL PVAc solution (indicated as P0.25 to P1.5). The structure of PEDOT-PSS is given in Fig. 4.2.

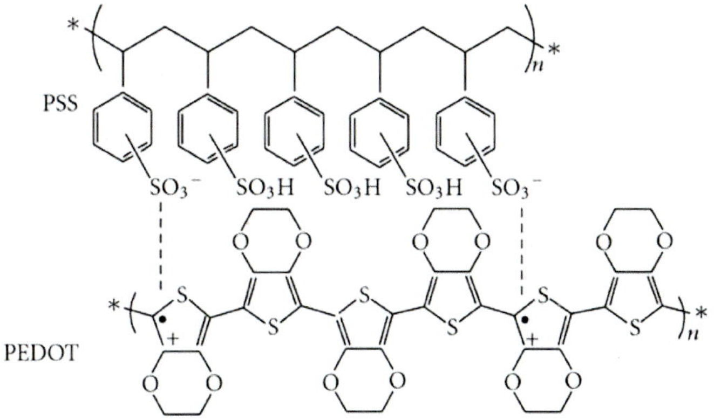

Figure 4.2 PEDOT-PSS structure.

The resulting solutions were stirrred magnetically at room temperature until homogenous solutions were obtained. The solutions of composites are shown in Fig. 4.3.

Figure 4.3 PVAc and PEDOT-PSS/PVAc solutions.

4.3.2 Synthesis of Poly(3,4-Ethylenedioxythiophene) in a Poly(Vinyl Acetate) Matrix

PEDOT was synthesized by chemical polymerization (Fig. 4.4) as follows: 1.0 g poly(vinyl acetate) (PVAc) was dissolved in 8 mL acetone and stirred magnetically at room temperature until a homogenous solution was obtained. An EDOT monomer was added dropwise to the PVAc solution. Then 25, 50, and 72 μL (2.35×10^{-4}, 4.7×10^{-4}, and 7.05×10^{-4} mmol) of EDOT were used for the reaction (S25, S50, and S75, respectively). To increase the yield of PEDOT, synthesis was carried out with excess oxidant 1:2 (EDOT: cerium(IV) ammonium nitrate (CAN)). A solution of CAN (respectively, 0.258, 0.515, and 0.773 mmol) dissolved in 2 mL acetone was then introduced to reaction mixture at once.

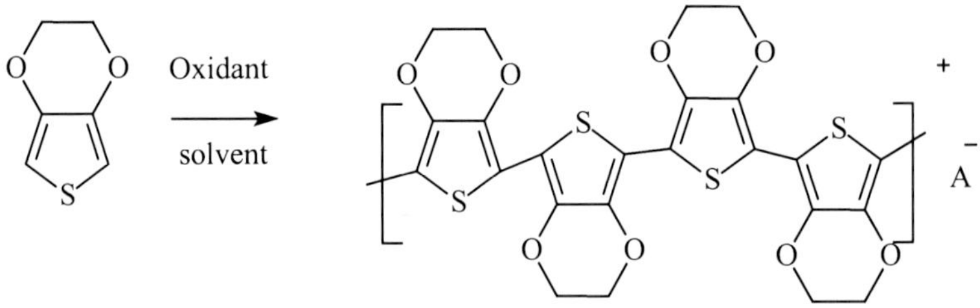

Oxidant
solvent
A^-

Figure 4.4 Doped PEDOT with anion of electrolyte after chemical polymerization.

The reaction mixtures were stirred for 24 hours at room temperature. The resulting dark-brown dispersions were used to prepare electrospinning solutions after adding DMF to the final solutions (details of characterizations of PEDOT/PVAC and PEDOT-PSS/PVAc systems are given in Chapters 5 and 6).

4.3.3 Preparation of Electrospinning Solutions: PEDOT-PSS in a PVAc Matrix

4.3.3.1 Electrospinning of PEDOT-PSS/PVAc

Series of PEDOT/PVAc and PEDOT-PSS/PVAc composites and 10 wt.% PVAc solution were prepared and stirred at room temperature with the speed of 400 rpm for 3 hrs. Electrospinning details are as follows[120]: The distance between the tip of the syringe needle and the Al plate collector was 15 cm and the flow rate of the solution was 0.5 mL/h, while the electrical potential between the needle tip and the Al plate was 10 kV.

Obtained fibers were characterized for both composites by Fourier transform infrared spectrophotometry–attenuated total reflectance (FTIR-ATR), UV-Vis spectrophotometer, electrochemical impedance spectroscopy, morphological analysis, and broad-band dielectric spectrometer.

4.3.3.1.1 *Process setup and electrospinning*

In the electrospinning process, the setup consisted of a direct current (DC) high-voltage power supply from GAMMA High Voltage Research Inc., USA (Model no: ES50) with an electrical potential range from 0 to 30 kV and a syringe pump (New Era Pump Systems Inc., USA Model no: NE-500). The metal collector was covered with an aluminum foil or indium tin oxide (ITO) polyethylene terephthalate (PET). The setup was kept in a plexiglass box for the experimenter's safety. All experiments were carried out under atmospheric pressure and at room temperature.

The positive electrode is the metal part of the needle and the negative part of the electrode is the metal collector. About 5 to 180 minutes of operation time was applied for the deposition of fibers on wax paper, aluminum foil, and ITO PET. A horizontal setup was chosen for the electrospinning process. A representative picture of design of electrospinning is illustrated in Fig. 4.5.

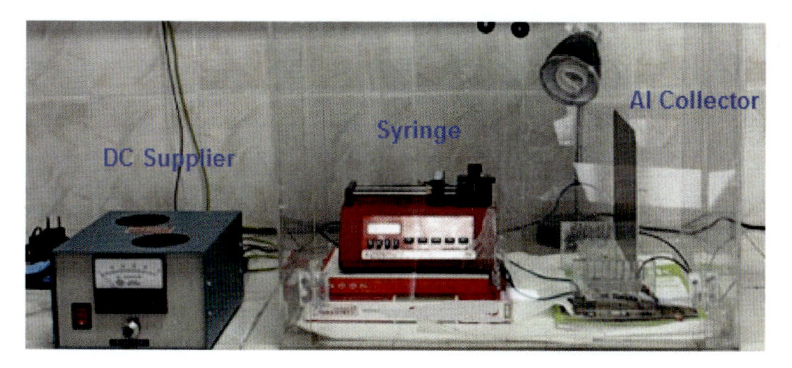

Figure 4.5 A representative picture of modular electrospinning units.

In some cases the angle of spinning (vertical position) to spin was chosen[121] vertically with a rotating collector (Fig. 4.6) and a speed of rotation of 1500 rpm and a voltage of 30 kV.

Some of the related parameters could be as follows:

- The diameter of the syringe or pipette: Three different pipette tip diameters were selected: D = 0.9 to 1.7 mm.

- The spinning height (distance from the tip of the pipette or syringe to the target): H = 10, 25, 40 cm, which reflects the charge density from 3 to 0.75 kV/cm.
- Spinneret, the radial distance of spinning: The distance from the center of the rotating disk to the center of pipette or syringe position was R = 4 to 14 cm.

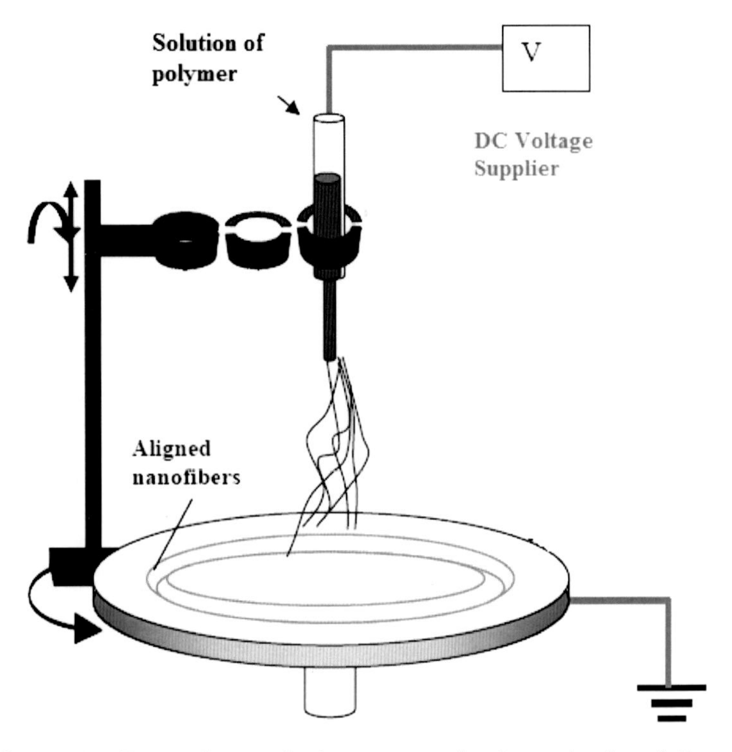

Figure 4.6 Rotary electrospinning apparatus for the production of aligned nanofibers.

4.4 Conjugated Polymeric Nanostructures

The typical conjugated polymers (CPs) with conjugated chain structures include polyacetylene, polypyrrole (PPy), polythiophene (PTh), polyaniline (PANI) and their derivatives.[122–124] The conjugated structures exhibit strong UV-Vis absorptions in visible region. The solutions or films of conjugated polymers are usually fluorescent and their fluorescence properties can be quenched by chain aggregation. Conjugated polymers in neutral state exhibit a small conductivity, in the range of about 10^{-10} to 10^{-5} S cm^{-1}, which can be considered as semiconductive, but at conductive states conductivities can be >1 S cm^{-1} by chemical or electrochemical "doping". After doping, the backbone of CP forms positive (p-doping) or negative charge carriers (n-doping), therefore, counter ions with opposite charges will be entrapped or released from the polymer matrix to maintain the charge neutrality of the polymer. As a result of doping, the volume of the polymer will be changed. Simultaneously, the absorption bands of the neutral conjugated polymers are reduced and new absorption bands associated with charge carriers will be formed at longer wavelengths.[125]

To improve the processibility of conjugated polymers and to improve performance of the organic polymeric electronic devices, CPs usually have to be nanostructured. Nanostructures can reduce the size of organic transistors and contribute to applications in thin panel displays. The solar cells fabricated by the use of the nanocomposites of poly(alkylthiophene) and a fullerene derivative revealed high efficiency in energy conversion due to the presence of heterojunctions in nanoscale. CP nanomaterials also have high specific surface areas which can accelerate interactions between CP and the analyte, by increasing the sensing response of the sensors based on CPs.[111,113,114]

CP nanomaterials can be synthesized by chemical or electrochemical methods. In chemical reactions generally powdery nanomaterials are produced and can be scaled up. Nanostructures of CP deposited on the electrode surface as films by electropolymerization have limited surfaces compared to the materials obtained by chemical synthesis. Nanostructures generally grow along the direction of the electric field to form oriented structures. The electrochemical polymerization reaction rate can be controlled through the applied potential or current density, and controlled amount of product can also be obtained. The morphology of the nanomaterials may also be modulated by the conditions of electropolymerization. Electrochemical polymerization is an effective technique for producing CP nanomaterials with controlled morphologies and properties.[125]

4.4.1 Electrochemical Polymerization of Heterocyclic and Aromatic Monomers

Thiophene, pyrrole, aniline, furan, etc. are used for electrosynthesizing CPs as aromatic and heterocyclic compounds, therefore these compounds can be oxidized to stable radical cations. The coupling of two radical cations or one radical and a monomer and successive removal of two protons results in the formation of a dimer. The widely accepted mechanism of thiophene dimer formation is shown in Fig 4.7, where an electrochemical process of oxidation and a chemical process of coupling and eliminating of protons are realized. Further oxidation of the dimers and coupling reactions produce oligomers, which are soluble in electrolytes and electroactive for polymerization. Further coupling reactions of oligomers through the same procedures result in the final polymers. Since more than two electrochemically active positions are mostly available for polymerization in the heterocyclic monomers, most electrosynthesized CPs have a cross-linked structure, and only the mostly preferred α,α'-coupling of thiophene is presented in Fig 4.7.[115,116] As the oligomer chains grow and polymer or the cross-linked structure is formed, CPs deposit at the working electrode surface. Oxidation of the polymer chains dopes the polymer into the conductive state. At the initial stage of electrochemical polymerization, the CP films are nearly smooth and compact, and their roughness and porosity increases as film thickness increases. The diffusion of counter ions into a compact film is more difficult than into a porous structure.[117] Accordingly, the doping level of a CP film synthesized potentiostatically increases during the electrolysis. CPs can be electrochemically synthesized by cyclic voltammetric, constant current and constant potential electrolysis. At constant current density conditions, the total amount of the CP can be controlled by the time of the electrolysis. The polymers produced at constant potential or current density are in the conductive (oxidized/doped) state. On the other hand, those prepared by cyclic voltammetric scanning depend on the potential at which the polymerization end.[125]

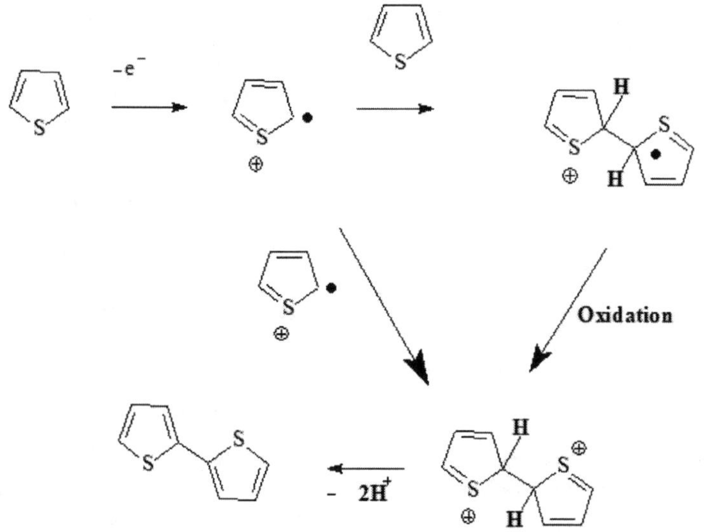

Figure 4.7 Initial step of electrochemical polymerization process of thiophene.

The oxidation potentials of aromatic compounds depend on their structures and the type of electrolytes. Electron donating groups or large dense aromatic rings can stabilize the corresponding radical cations by reducing the oxidation potentials, i.e., the oxidation potential of 3-methylthiophene is lower than that of thiophene, while that of halides (in 3rd position of ring) containing thiophenes is higher than that of thiophene. The oxidation potential of naphthalene or pyrene is quite lower than that of benzene. Some electrolyte molecules have electrocatalytic properties, by activating the monomers and lowering their oxidation potentials, i.e., the oxidation potential of thiophene in boron trifluoride diethyl etherate (BFEE) was measured to be 1.2 V (vs. SCE), which is lower than that measured in a neutral electrolyte acetonitrile (2.1 V, vs. SCE).

The heterocyclic monomers, i.e., pyrrole, aniline and 3, 4-ethylenedioxythiophene (EDOT), with oxidation potentials lower than that of water decomposition can be electrochemically polymerized in aqueous media. However, the solubility of aromatic compounds in water is mostly poor and the polymerization rates are low. Acidic medium improves the solubility of pyrrole and aniline and the ionic conductivity of electrolyte. The use of surfactant as an addition agent can also improve the solubility of monomers. Electrochemical polymerizations are generally performed in organic media, due to the good solubility of the monomers. Since the electropolymerization reaction proceeds by radical cation intermediates, aprotic solvents (acetonitrile, benzonitrile, dichloromethane, etc.) are used. Different ionic liquids are also used as the solvents for electrochemical polymerizations.[126,127]

Electrochemical polymerizations are carried out in a 3-electrode electrochemical cell, consisting a working electrode (WE), a counter electrode (CE), and a reference electrode (RE). To reduce the polarization the size of the counter electrode should be same or larger than the working electrode. As working electrodes metals (i.e., Pt, Au, Ni), alloys (i.e., stainless steel), and glassy carbon and conductive oxide (such as indium tin oxide (ITO)) electrodes are used. The counter electrode can be the same or different to the working electrode, it should be chemically and electrochemically inert to monomers, solvents, and supporting electrolytes. The saturated calomel electrode (SCE) has been used as a reference electrode in aqueous or organic media. Ag/AgCl or Ag wire is also utilized as a pseudo reference electrode by directly immersing it in the electrolytes.

4.4.2 Composites with Carbon Nanomaterials

Aligned arrays of carbon nanotubes (CNTs) have advantages to be used in many applications,[128] i.e., electron-emitting flat-panel displays and electrochemical actuators. So they can be used as the working electrode for the electrodeposition of CPs. The CNT electrode can be prepared by gold sputter coating film onto the amorphous carbon layer covered on an as-synthesized aligned CNT film and then separating the gold-covered aligned CNT film from the quartz glass plate with an aqueous HF solution. Alternatively, aligned CNT nanowires were obtained electrochemically by the oxidation of the corresponding monomer at the CNT electrode (Fig. 4.8a).[129]

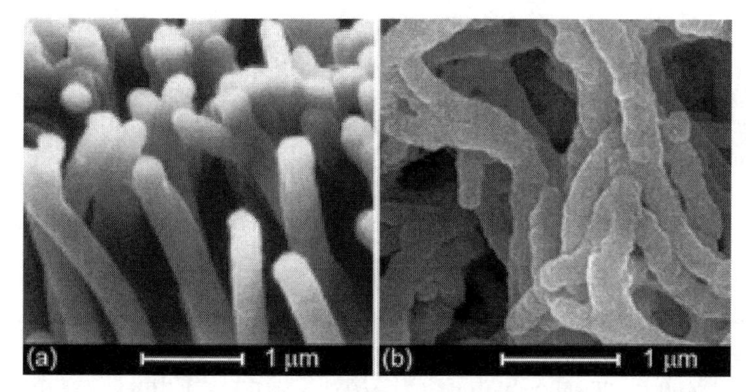

Figure 4.8 SEM images of CP/CNT nanocomposites. (a) Aligned PPy/CNT nanocomposite. Reprinted with permission from Ref. 128. Copyright © 2000 WILEY-VCH Verlag GmbH, Weinheim, Fed. Rep. of Germany. (b) Oligopyrene/SWCNT nanocomposite. Reprinted with permission from Ref. 131. Copyright © 2005, Royal Society of Chemistry.

CP/CNT nanocomposites can be prepared by the electrochemical codeposition of solvent-dispersed CNTs into CP film through electrostatic interaction.[130] By this method, nanocomposite films of CNT with PPy, PANI and PEDOT were prepared, in which CNTs bearing carboxylate groups, surfactant-stabilized or DNA-wrapped CNTs with negative charges acted as both 1D templates and dopants for growing CPs. Noncovalent interactions between CP aromatic or heterocyclic rings and basal graphene planes of single walled CNTs are also used as the driving force for coating CP layers onto CNTs. For example, pyrenebutyric acid dispersed SWCNTs in BFEE were electropolymerized to prepare oligopyrene/SWCNT nanocomposites (Fig. 4.8b).[131]

4.4.3 Composites with Insulating Polymers

Physical and chemical limitations of CP can be overcome by the use of insulating polymer as a matrix, including the processability and mechanical and thermochemical stability of CPs. The resulting composite incorporate the advantages of electrical, redox, or optical properties of CPs with the mechanical properties of the insulating polymer. Insulating polymers have been generally used as templates for the electrosynthesis of CP/insulating polymer nanocomposites. Nanofibers of poly(acrylonitrile-co-methylacrylate)/polypyrrole core–shell nanoparticles, and polypyrrole/poly(acrylonitrile-co-butyl acrylate) nanoparticles are homogeneously prepared by micro emulsion polymerization.[132,133] DNA chains immobilized on the electrode surface as the template for electrochemical polymerization of aniline. Electrostatic interaction between DNA and aniline allow the growth of PANI along the DNA strands.

Multilayers of PS colloidal crystals assembled on the electrode surface were also used as the template for electropolymerization of aniline.[130] Nanoporous or nanocomposite films of PANI may be produced with or without the removal of the PS template. Better stability and reproducibility of the nanocomposite film have been observed than the nanoporous PANI film. The reason for this was the PS nanoparticles enable the PANI/PS composite film to maintain its morphology, since the nanoporous PANI film may suffer from collapse or shrinkage. Insulating polymer nanoparticles can also be electrochemically codeposited into the CP matrix by grafting their surfaces with reactive moieties,[134] i.e., a PPy-PEG-PLA nanocomposite film can be prepared. [PEG-PLA:poly(1-ethoxyethylglycidyl ether)-block-poly(L-lactide) copolymer].

An electrohydrodynamic polymerization technique was introduced for the electrosynthesis of CP/insulating polymer nanocomposites. By this method optically active PANI colloids were prepared in a divided electrochemical flow-through cell via the polymerization of aniline in aqueous sulfuric acid/enantiomer of 10-camphorsulfonic acid, and polystyrenesulfonic acid as a stabilizer and codopant.[135]

4.4.4 Composites with Metal or Oxide Nanoparticles

Hybrid organic–inorganic materials based on CPs have been studied, inorganic nanoparticles with different properties and sizes are blended with CPs, resulting in nanocomposites with specific properties acquired from individual components. For instance, the magnetic susceptibility of γ-Fe_2O_3, the electrochromic property of WO_3 and the catalytic activity of Pd, Pt, Au, etc. have been incorporated with CPs in the resulting nanocomposites.

Electrochemical codeposition is used for the incorporation of negatively charged WO_3, SiO_2, SnO_2 and Ta_2O_5 particles into an electropolymerized PPy film by controlling the pH of aqueous dispersions of pyrrole and the corresponding particles.[136]

To create negative charges on particle surfaces, the pH of the electrolyte should be adjusted to above the isoelectric point of the respective oxide. One-step electrochemical codeposition method was proposed for the synthesis of Ag/PANI composite nanowires.[137]

Metal oxide nanoparticles can be integrated into CP matrices by using monomer-capped nanoparticles as a starting component. The monomer on the nanoparticle surfaces can be electropolymerized with monomers dissolved in solution to form a nanocomposite film, i.e., pyrrole-capped TiO_2 nanoparticles were produced by grafting a bifunctional ligand with pyrrole and acetylacetone groups onto TiO_2 nanoparticle surfaces. Electrolysis of modified nanoparticles generate PPy/TiO_2 nanocomposite film.[138] Gold nanoparticles stabilized by a monolayer of 5-(3-thienyl)-1-pentanethiol (TPT), and later they were incorporated into poly(3-alkylthiophene) nanowires through the electro-copolymerization of the thiophene moieties of TPT ligands and 3-butylthiophene in anodic alumunium oxide (AAO) template.[139]

Metal nanoparticles can also be electrodeposited onto a CP modified electrode to obtain nanocomposites. The number, size and distribution of metal particles can be controlled by the conditions of electrochemical deposition.[140] The pulse potentiostatic technique results in an efficient dispersion of Pt particles in the polymer matrix.

4.4.5 Applications of Electrosynthesized CP Nanomaterials

CPs and their composites are utilized in the fields of electrochemistry, electroanalysis, electrocatalysis, batteries and capacitors, etc as electrode. In addition to the conductivity and electroactivity of CPs, small ions and molecules can diffuse into the CP matrices, providing further improvement compared to the conventional electrode materials. Efficiently using all the active sites and enhancing mass transport during the electrode process, the thickness of the CP film can be reduced to allow the ion diffusion in the CP matrices. By these properties CP nanomaterials exhibit better performances, due to their larger specific surface areas and small dimensions. Additionally, nanostructures of CPs may produce new surface properties and better functionalities.

4.4.5.1 Sensors

CPs are convenient matrices for the immobilization of enzymes and other biological molecules,[141] forming different selective electrodes.

CP modified electrodes are used as chemo- and bio-sensors in recent years.[142] Electropolymerization is a convenient method to synthesize a thin CP film. Moreover, the immobilization of the enzyme can be attained by an in situ electro-codeposition. PPy/ CNT/ glucose oxidase (GOx) modified electrode was prepared by electro-codeposition of PPy and GOx onto a CNT array, indicating high sensitivity towards glucose, about one order of magnitude higher than that of the PPy film grown on a gold electrode.[143] This increase in sensitivity can be attributed to the larger surface area, which allows glucose to easily access GOx loaded in the thin PPy film.

Electrosynthesized CP nanostructures might be used as the gas sensing layers of chemiresistors. Gas sensors built on single CP nanowires can be fabricated by combination of electrosynthesis and the micro-process technique. The improved performances of gas sensors are generally due to the high specific surface areas of the nanostructures, by allowing the analyte molecules easily access to the active sites of CP matrices. The presence of further elements of CP nanocomposites may induce unique electric properties or catalytic activities towards the analytes, further improving the performances of the sensors.

4.4.5.2 Fuel cell electrode

Due to its low cost and high energy density, the methanol fuel cell is one of the substantial fuel cell. Oxidation of methanol is generally difficult and noble metal nanoparticles catalysts are needed. To have a uniform dispersion and good stability of catalyst, metal nanoparticles have to be properly deposited on a supporting substrate. The extensively used catalyst supports for fuel cell electrodes are different carbon materials,[144] on the other hand, processing and fabrication of carbon electrodes are difficult. Accordingly, CPs were developed as alternative supports for the dispersion of metal particle catalysts, the CP/metal nanoparticle composite electrodes can load more catalysts as a thin film. For electro-oxidation of methanol, PPy/ Pt and PANI/Pt nanocomposites as the catalytic electrodes have been well studied, and recently, research has been concentrated on shaping the CP matrices into a nanostructured morphology.[145] For instance, Pt nanoparticles were electrochemically deposited in an electrodeposited PPy nanowire matrix, forming a PPy/Pt nanocomposite. The PPy/Pt composite modified electrode exhibited good catalytic activities concerning the oxygen reduction and the methanol oxidation in comparison with the Pt electrode.

4.4.5.3 Batteries

CPs have generally been investigated as a cathodic material in combination with a metallic anode. There are some difficulty of usage of CP cathodes, including poor cyclability and low conductivity in the reduced state. For this reason, CPs were frequently incorporated with inorganic metal oxides or CNTs, forming nanostructured hybrid electrode materials. Greater number of the hybrid electrode materials were prepared by in situ chemical polymerization. Electrochemical polymerization was used to produce nanocomposite arrays on the electrode surfaces, considered as good electrode materials, i.e., PANI/CNT composite electrodes were synthesized by electropolymerization on CNT. The presence of CNT increases the surface area and provide the conductivity of the composite as PANI is reduced. Accordingly, the lithium battery assembled with a PANI/CNT cathode and gel polymer electrolyte showed a high energy density of 86 mAh g^{-1}. PANI was also electrochemically deposited onto Ni foam supported porous NiO film, constructing a NiO/PANI nanocomposite electrode. The PANI component enhanced the electrical conductivity and film stability, resulting less polarization and good cycling performance. The specific capacity after 50 cycles for NiO/PANI film was found to be 520 mAh g^{-1} at 1°C, being higher than that of the NiO film (440 mAh g^{-1}).[146]

4.4.5.4 Electrochemical actuators

Conjugated polymers have been used extensively for the fabrication of electrochemical soft actuators. By the redox processes, electrochemo-mechanical deformation of CP films is affected by the film morphology, and construction type of the polymeric film layers. Most of the CP electrochemical actuators have been constructed based on bilayer or multilayer configuration. Monolithic PPy film (and by the inclusion of small amount of oxide nanoparticles) was also used to produce actuators, by exhibiting a large bending angle, with a high response rate and a long lifetime. An actuator consisting of a PPy nanowire and a PPy thin film can capture PS microparticles from a dilute aqueous dispersion and transfer the particles to a new medium by sonication or electrochemical control. An actuator with an inverse opal structure can selectively trap particles with sizes smaller than the orifices of its surface caves.[147] So, it can be used to separate different sized nanoparticles from their mixed dispersion. But through a sandwich structure design, linear actuation can be converted to bending actuation. A typical structure contains two

PPy layers that forms the outside layers on either side of a polymer electrolyte layer in the middle. The polymer electrolyte layer could be solid state electrolyte or electrolyte saturated media (i.e., porous PVDF film). Once a voltage is applied, one PPy layer performs as cathode and the other performs as the anode. Since oxidation and reduction occur separately at both sides, expansion and contraction on each side will cause bending actuation (Fig 4.9).

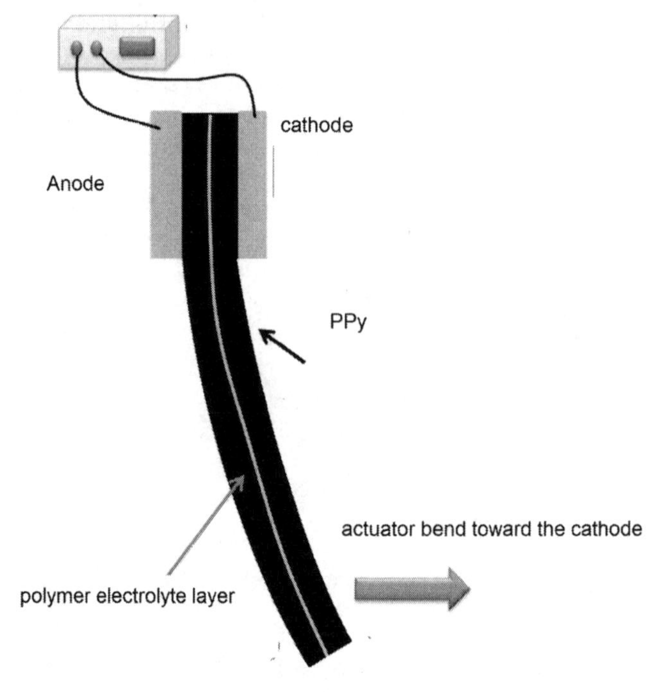

Figure 4.9 Schematic structure of PPy bender and its working state.

4.5 MWCNT/PS and PPy/PS Nanofibers

Poor processibility (dissolving in a solvent or melted by heat) results in difficulties in electrospinning conductive polymers on their addition of a spinnable polymer. The addition of conductive particles such as CNTs in a polymer matrix, starting from polymer precursors and converting them into conductive polymer in the later step, and using a core–shell coaxial electrospinning strategy and removing the nonconductive core or shell is a method used for obtaining different nanofiber formations. CNTs are used as conductive fillers in electrospun fibers due to their high electrical and thermal conductivities and mechanical properties.

Multiwalled carbon nanotube (MWCNT)/PS solution blends were prepared by adding 1% to 10% (wt.% to PS) of MWCNTs to DMF.[149] The MWCNT/DMF mixture was mechanically stirred for half an hour. Then 25% (wt.% to solution) PS was added to this mixture. To prepare for $FeCl_3$/PS solutions, 40% or 60% (wt.% to PS) of $FeCl_3$ were first added to DMF and stirred until $FeCl_3$ was fully dissolved (~10 min). A 15% (wt.% to solution) PS was then added to the mixture. All polymer solutions were stirred at 50°C overnight in a closed glass vial. The solutions were cooled to room temperature before electrospinning. Spinning of the MWCNT/PS fibers was done at room temperature (23 ± 2°C) and the RH was kept at 19 ± 1% RH. The distance of the aluminum collector from the tip of the needle was 10 cm. Conductivity increase and fiber diameter change was followed by the increase of MWCNTs (Figs. 4.10 and 4.11).[148]

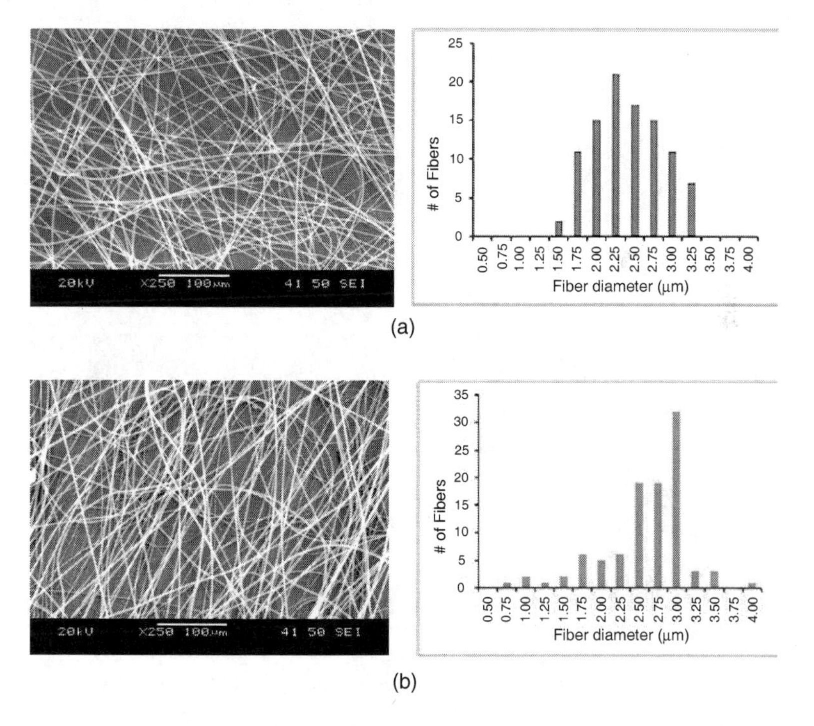

Figure 4.10 SEM images and fiber diameter distribution histogram of electrospun MWCNT/PS fibers with various MWCNT concentrations: (a) pure PS and (b) 1% MWCNT/PS. Reprinted from Ref. 148 with permission from SPIE.

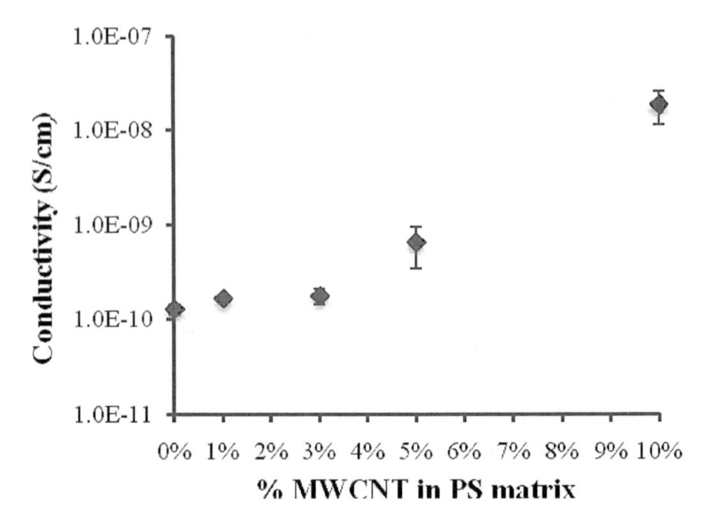

Figure 4.11 Two-point electrical conductivity measurements of MWCNT/PS electrospun fibrous films as a function of MWCNT concentration. Reprinted from Ref. 148 with permission from SPIE.

When the electrospun oxidant fibers are exposed to Py monomer vapors, redox reaction occurs where Py is oxidized, while $FeCl_3$ is reduced. During the Py vapor diffusion, which adds moisture to the fibers, the longer the polymerization time, the more compact the fibers (Fig. 4.12).

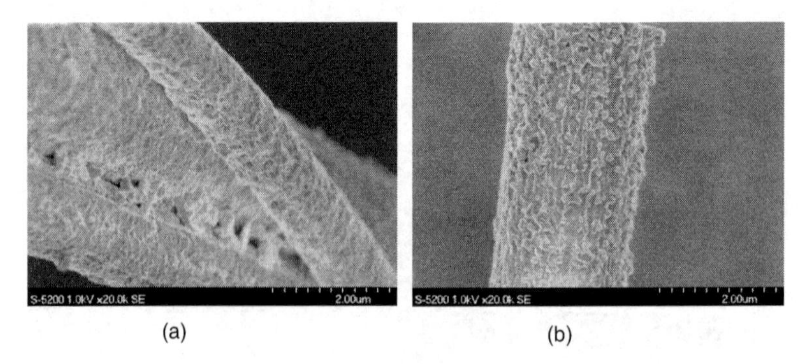

Figure 4.12 SEM images of (a) PPy-coated PS fibers (by $FeCl_3$) at 20k× magnification for 60 min polymerization time and (b) 100 min. Reprinted from Ref. 148 with permission from SPIE.

Chapter 5

Electrospinning: The Velocity Profile

5.1 Electrospinning: The Velocity Profile

Taylor identified the critical electric potential for electrostatically forming a cone of liquid (Taylor cone-semivertical sphere) at the end of a capillary tube. The derivation began with the expression for the equilibrium state of a droplet at the end of a pressurized tube, and the coefficients for the electrostatic potential were generated by observing the deflection of charged solutions at the end of an inverted capillary. Taylor showed that the vertical voltage (V_c in kV) at which the maximum jet fluid instability start, is given by[9]

$$V_c^2 = 4\frac{H^2}{L^2}\left[\ln\frac{2L}{R} - 1.5\right](0.117\pi.R.\gamma)$$

where H is the distance between the electrodes (the capillary tip and the collecting screen), L is the length of the capillary tube, R is the radius of the tube, and γ is the surface tension of the fluid (units: H, L, and R in cm; γ in dyne per cm). In spinning, the flow beyond the spinneret is mainly elongational. The minimum spraying potential of a suspended, hemispherical, conducting drop in air is

$$V = 300\sqrt{20\pi.r.\gamma}$$

Nanofibers of Conjugated Polymers
A. Sezai Sarac

ISBN 978-981-4613-51-4 (Hardcover), 978-981-4613-52-1 (eBook)
www.panstanford.com

where r is the jet radius. If the surrounding medium is not air but a nonconductive liquid immiscible with the spinning fluid, drop distortion will be greater at any given electric field and, therefore, the minimum spinning voltage will be reduced. If the electrospinning process is realized in vacuum, the required voltage will lower.

The spinning rate was estimated by measuring the length of the polyethylene (PE) fiber obtained in a known interval of time.[121] The rate was of the order of 1 m/min, and it increased with the applied field strength. The relationship between the spinning rate and electric field strength E is found to be as following at 200°C.

$$\text{Rate} = Ae^{bE}$$

where A and b are constants with values 38.7 and 0.16, respectively, if the rate is expressed in cm/min and E in kV/cm.

The relationship between the capillary radius plane distance and the applied voltage is given by the following equation:

$$V_{\text{o}} = A_1 \left[\frac{2Tr\cos\theta_{\text{o}}}{\varepsilon_{\text{o}}} \right]^{0.5} . \ln \left[\frac{4h}{r_{\text{c}}} \right]$$

where T, θ_{o}, and ε_{o} are surface tension, cone half angle, and permittivity of free space, respectively. It was concluded that the capillary radius would be increased as the onset potential increased, the plane distance would increase with increasing onset potential, and the surface tension would be in direct proportion to the onset potential.

The velocity profile inside the liquid cone at the base of an electrically driven jet was examined by inserting tracer particles into the liquid.[149] Observations of these particles at high magnification demonstrated the presence of an axisymmetric circulation inside the cone due to interfacial electrical shear stress. An analytical solution was presented that predicts the velocity profile inside the cone. The surface velocity were expressed as

$$v_{\text{r}} = \frac{1}{r}\left(-\boldsymbol{B} + 2\boldsymbol{C}\cos\theta_{\text{o}}\right) + \frac{\boldsymbol{v}}{2\pi(1-\cos\theta_{\text{o}})r^2}$$

where $\boldsymbol{B}$ and $\boldsymbol{C}$ are arbitrary constants, $\boldsymbol{\theta}$ is the semicone angle, $\boldsymbol{r}$ is the spherical radius of the cone, $\boldsymbol{v}$ is the volumetric flow rate, and $2\pi(1 - \cos\boldsymbol{\theta}_{\text{o}})\boldsymbol{r}^2$ is the steradian area of the cone.

A relationship between the applied voltage, surface tension, air permittivity, and capillary radius was also introduced. The difference between the calculated jet diameter from the equation and the experimentally measured value had been recorded.

$$E = \sqrt{\frac{4\gamma}{\varepsilon_0 R}}$$

where E is the electrospinning voltage, R is the capillary radius, and γ is the surface tension.

The reasons for the instability were explained by using a mathematical model. The rheological complexity of the polymer solution was included, which allowed consideration of viscoelastic jets. It was shown that the longitudinal stress caused by the external electric field acting on the charge carried by the jet stabilized the straight jet for some distance. In response to the repulsive forces between adjacent elements of charge carried by the jet, a lateral perturbation was observed, and the motion of segments of the jet changed rapidly into an electrically driven bending instability. The 3D paths of continuous jets were calculated, both in the nearly straight region where the instability grew slowly and in the region where the bending dominated the path of the jet.[150]

Recent studies, have demonstrated that the key role in reducing the jet diameter from a micrometer to a nanometer is played by a nonaxisymmetric or whipping instability, which causes bending and stretching of the jet in very high frequencies. Shin et al. investigated the stability of electrospinning PEO (polyethylene oxide) jet using a technique of asymptotic expansion for the equations of electrohydrodynamics in powers of the aspect ratio of the perturbation quantity, which was the radius of the jet and was assumed to be small. After solving the governing equations it was found that the possibility for three types of instabilities exists. The first instability is the classical Rayleigh instability and is axisymmetric with respect to the centerline of the jet. The second is again an axisymmetric instability and may be referred to as the second axisymmetric instability. The third is a nonaxisymmetric instability, called "whipping" instability, mainly by the bending force. Keeping all the other parameters unchanged, the electric field strength will be proportional to the instability level. Namely, when the field is the lowest, the Rayleigh instability occurs, whereas the bending (or whipping) instability corresponds to the highest field It was experimentally observed that the phenomenon of so-called "inverse cone" in which the primary jet was thought to be split into multiple jets is actually caused by the bending instability. At higher resolution and with shorter electronic camera exposure time, the inverse cone is not due to splitting but as a consequence of small lateral fluctuations in the centerline of the jet.[9]

Related to the fluid jet diameter, as viscosity of the polymer solvent increased, the spinning drop changed from approximately hemispherical to conical. By using equipotential line approximation calculation, the radius r_0 of a spherical drop (jet) was calculated, as follows:

$$r_0^3 = \frac{4\varepsilon \dot{m}_0}{k\pi\sigma\rho}$$

where ε is the permittivity of the fluid (in coulombs/volt-cm), $\dot{m}_0$ is the mass flow rate (g/sec) at the moment when r_0 is to be calculated, k is a dimensionless parameter related to the electric current, σ is the electric conductivity (A/volt-cm), and ρ is density (g/cm^3).

A general electrohydrodynamic model of a weakly conductive viscous jet accelerated by an external electric field was also derived, by considering inertial, hydrostatic, viscous, electric, and surface tension forces. Nonlinear rheologic constitutive equation for the jet radius was derived,

$$R^* = 2\left[\varepsilon_0 \sigma_s\right]^{0.33} \left(\frac{Q}{J}\right)^{0.66}.$$

where σ_s is the coefficient of surface tension, ε_0 is the permittivity of vacuum, Q is the volumetric flow rate, and J is the electric current. The mathematical model successfully predicted experimental observations, which could be used for better control and optimization of the electrospinning process.

Experimental results indicate that the diameter of the fiber produced by electrospinning is influenced by the polymer concentration and molecular conformation,such influence is, described by the Berry number, B_e, on an electrospun poly (L-lactic acid)/chloroform system gives the following equation:

$$B_e = [\eta]C$$

where $[\eta]$ is the intrinsic viscosity of the polymer ($[\eta]$ = the ratio of specific viscosity to concentration at infinite dilution) and C is the concentration of the solution. The degree of entanglement of polymer chains in solution could be described by B_e. When the polymer is in a very dilute solution, polymer molecules are so far apart in the solution that polymer molecules are sparsely distributed and B_e is less than unity. There is a less possibility of individual molecules entangle with each other. As the polymer concentration is increased, the level of molecular entanglement increases, and B_e becomes greater than unity.

5.2 Electrospinning and Nanofibers: Applications

One of the most important applications of traditional (micro-size) fibers, especially engineering fibers such as carbon, glass, and Kevlar fibers, is to be used as reinforcements in composite developments. With these reinforcements, the composite materials can provide superior structural properties such as high modulus and strength to weight ratios, which generally cannot be achieved by other engineered monolithic materials alone. Nanofibers will also eventually find important applications in making nanocomposites. This is because nanofibers can have even better mechanical properties than micro fibers of the same materials, and hence the superior structural properties of nanocomposites can be anticipated. Moreover, nanofiber reinforced composites may possess some additional merits which cannot be shared by traditional (microfiber) composites. For instance, if there is a difference in refractive indices between fiber and matrix, the resulting composite becomes opaque or nontransparent due to light scattering. This limitation, however, can be taken care of by making the fiber diameters significantly smaller than the wavelength of visible light. In addition to composite reinforcement, other application fields based on electrospun polymer nanofibers have grown steadily, especially in recent years.[9]

More extended or perspective application areas are summarized in Fig. 5.1.

Thin fibers of calf thymus Na-DNA were electrospun from aqueous solutions with concentrations from 0.3% to 1.5%. The electrospun DNA fibers had diameters around 50 to 80 nm. During electrospinning a process called splaying caused the jet to split longitudinally into two smaller jets, which split again, repeatedly, until the very small-diameter fibers were formed. The small-diameter fibers were transparent in ordinary 100 kV electron microscopes. Fibers could be spun from samples of DNA as small as 1 mg.

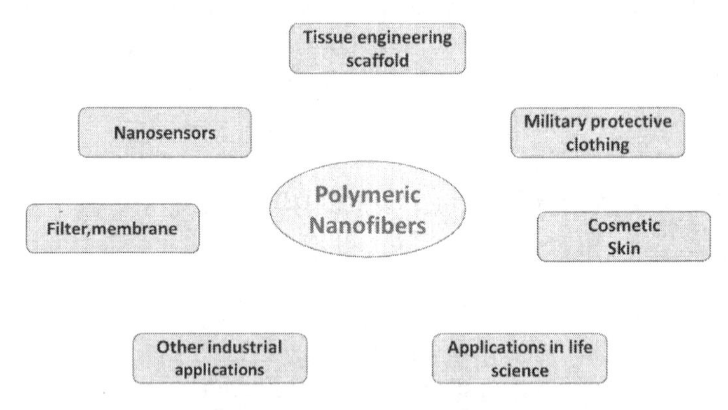

Figure 5.1 Some potential applications of electrospun polymer nanofibers.

Electrospun carbon precursor fibers, based on polyacrylonitrile (PAN) and mesophase pitch, having diameters in the range from 100 nm to a few microns, were stabilized and carbonized. These carbon nanofibers had a very high aspect ratio. Nanopores were produced in CNFs made from PAN by a high-temperature reaction with water vapor carried in nitrogen gas by increasing the surface area per unit mass of carbon black. For conductive CNT/polymer composite fibers, CNTs were incorporated into poly(vinylidene fluoride) (PVDF) in *N,N*-dimethylformamide (DMF) solutions and electrospun to form CNT/PVDF fiber mats.The thinnest fiber was obtained as 70 nm in diameter.

The reinforcing effects of nanofibers in an epoxy matrix and in a rubber matrix using electrospun nanofibers of polybenzimidazole (PBI) was observed. The average diameter of the electrospun fibers was around 300 nm. The nanofibers toughened the brittle epoxy resin. The fracture toughness and modulus of the nanofiber (15 wt.%)-reinforced epoxy composite were both higher than for an epoxy composite made with PBI fibrils (17 wt.%), which were whisker-like particles.

A shish-kebab model for the filament morphology was proposed. The electrospinning process was shown to be a means of creating porous thin films with structural gradients and controlled morphology that could enhance biocompatibility.

> Electrospinning a polymer solution could produce thin fibers with a variety of cross-sectional shapes. Branched fibers, flat ribbons, ribbons with other shapes, and fibers that were split longitudinally from larger fibers were observed. The transverse dimensions of these asymmetric fibers were typically 1000 to 2000 nm, measured in the widest direction. The observation of fibers with these cross-sectional shapes from a number of different kinds of polymers and solvents indicated that fluid mechanical effects, electrical charge carried with the jet, and evaporation of the solvent all contributed to the formation of the fibers.*

> The influences of surfactants and medical drugs on the diameter size and uniformity of electrospun poly(L-lactic acid) (PLLA) fibers by adding various surfactants (cationic, anionic, and nonionic) and typical drugs into the PLLA solution were investigated, significant diameter reduction and uniformity improvement were observed. It was shown that the drugs were capsulated inside of the fibers and the drug release in the presence of protein K followed nearly zero-order kinetics due to the degradation of the PLLA fibers.**

Polymer nanofibers can also be used for the treatment of wounds or burns of human skin, as well as designed for hemostatic devices with some unique characteristics. With the aid of an electric field, fine fibers of biodegradable polymers can be directly sprayed/spun onto the injured location of the skin to form a fibrous mat dressing, which can let wounds heal by enhancing normal skin growth and eliminating the formation of scar tissue, which would occur in a traditional treatment.

> An electrospinning process was used to fabricate silk fibroin (SF) nanofiber nonwovens for wound dressing applications. The electrospinning of regenerated silk fibroin (SF) was performed with formic acid as a spinning solvent. For crystallization, as-spun SF nanofiber nonwovens were chemically treated with an

---

*Excerpt reprinted with permission from Ref. 151, Copyright 2001, John Wiley & Sons.

**Excerpt reprinted with permission from Ref. 152, Copyright 2003, Elsevier.

aqueous methanol solution of 50%. The morphology, porosity and conformational structures of as-spun and chemically treated SF nanofibers were investigated by scanning electron microscopy (SEM), mercury porosimetry, wide angle X-Ray diffraction (WAXD), attenuated total reflectance infrared spectroscopy (ATR-IR), solid state ^{13}C CP/MAS nuclear magnetic resonance (NMR) spectroscopy. SEM micrograph showed that the electrospun SF nanofibers had an average diameter of 80 nm and a distribution in diameter ranging from 30 to 120 nm. The porosity of as-spun SF nanofiber nonwovens was 76.1%, indicating it was highly porous. Conformational transitions of the as-spun SF nanofibers from random coil to β-sheet by aqueous methanol treatment occurred rapidly within 10 min, confirmed by solid-state ^{13}C NMR, ATR-IR, and X-Ray diffraction.*

Favorable interactions between biodegradable polymer nanofibrous scaffolds with aligned architecture via electrospinning and the unique scaffold as well as a directional growth of the cells along the fiber orientation were demonstrated by cell morphology, adhesion, and proliferation studies. These results suggest a huge potential of this as a scaffold for blood vessel engineering.

5.3 Electrospinning of Polyaniline Blends

Nanofibers of polyaniline doped with camphorsulfonic acid (PANI.CSA) blended with PEO were prepared by the electrospinning technique. The morphology and fiber diameter of electrospun PANI blend fibers revealed that both the PEO and PANI.CSA/PEO blend fibers had a diameter ranging between 950 nm and 2100 nm, with a generally uniform thickness along the fiber. The conductivity of PANI-CSA/PEO electrospun fiber mat was found to be lower than that for a cast film due to the porous nature of fiber, but have shown similar Uv-vis.absorption characteristics. Although the conductivity of conducting nanofiber is relatively low compared to cast film, their porous structure and high surface to volume ratio enable faster doping. The rate for the vapor-phase dedoping/redoping of the electrospun fibers was at least 1 order of magnitude faster than for cast films. Long nanofibers of conducting electronic polymer blends with conventional polymers were conveniently fabricated in air by electrospinning.

*Excerpt reprinted with permission from Ref. 153, Copyright 2003, Nature Publishing Group.

Coaxial spinning can be used in producing PANI nanofibers.[154] This kind of electrospinning uses two spinnerets, allowing the low elastic fluid to elongate along with the electrospinnable fluid. This process will result in nanofiber with continuous core-sheath morphology. PANI nanofibers are produced with a dopant of CSA blended with poly(methyl methacrylate) (PMMA) in core-sheath form through coaxial electrospinning (Fig. 5.2). 100% doped-PANI fibers were then obtained by immersing the fiber blend into isopropyl alcohol. This was done to remove the PMMA shells and thereby release the doped PANI cores. Due to the removal, the diameters of the fibers decreased from 1440 to 620 nm.

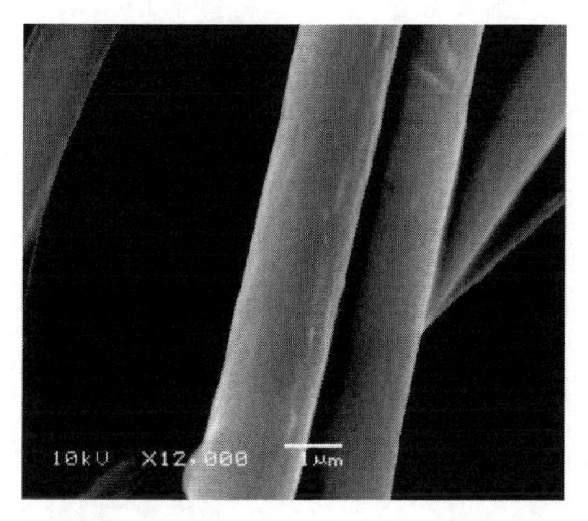

Figure 5.2 SEM images of electrospun PANI–PMMA core–shell fibers. Reprinted with permission from Ref. 154, Copyright © 2012, American Chemical Society.

The effects of polymer blend composition, degree of dilution, and collecting distance on fiber diameter and morphology of porous fiber webs for various diluted PANI/PEO solutions in an electrospinning process was investigated. Electrical conductivity of polyaniline fiber and mats as-electrospun, doped with an equimolar amount of CSA as a function of the PANI weight fraction in the blended fibers are given in Figs. 5.3 and 5.4.

It was found that distance was not a major factor on the diameter of nanofibers; rather, it was found to be a sensitive function of solution viscosity where fiber diameter decreased as viscosity decreased.[154] CSA-doped polyaniline (PANI)/PEO composite nanofibers with different compositions (12 to 52 wt.% of PANI) were synthesized

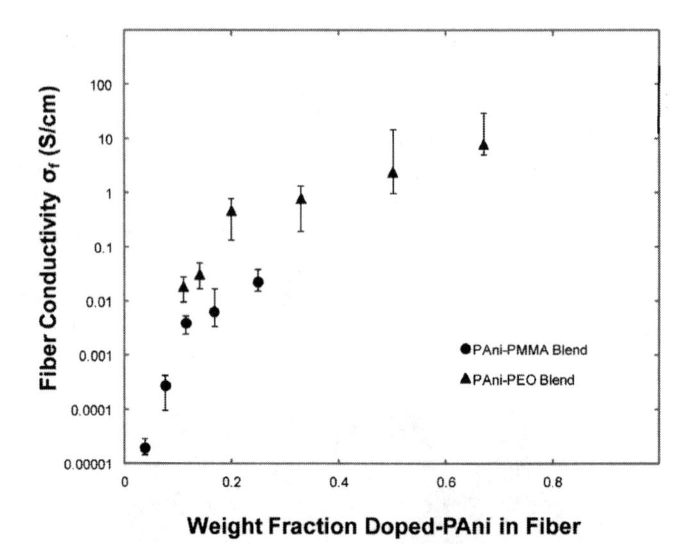

Figure 5.3 Electrical conductivity of as-electrospun polyaniline fibers (nominally doped with an equimolar amount of CSA) as a function of the weight fraction of PANI in the blended fibers. Reprinted with permission from Ref. 154, Copyright © 2012, American Chemical Society.

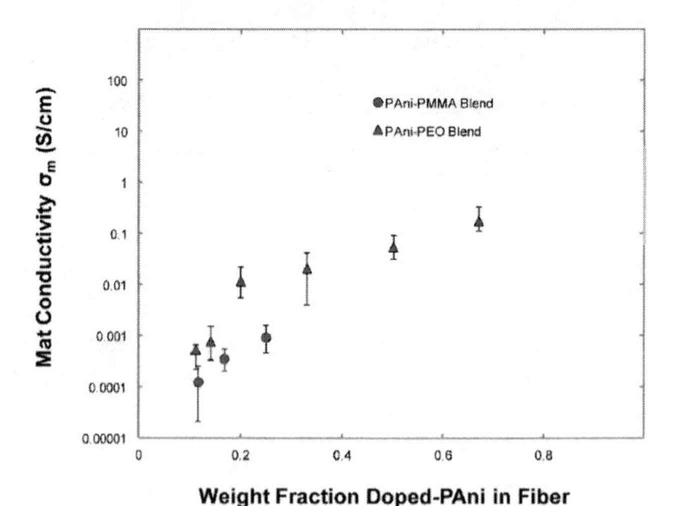

Figure 5.4 Mat electrical conductivity of polyaniline fiber mats (as-electrospun, nominally doped with an equimolar amount of CSA) as a function of the PANI weight fraction in the blended fibers. Reprinted with permission from Ref. 154, Copyright © 2012, American Chemical Society.

by an electrospinning method and their properties including optical, electrical and sensing were investigated (Figs. 5.5 and 5.6). The

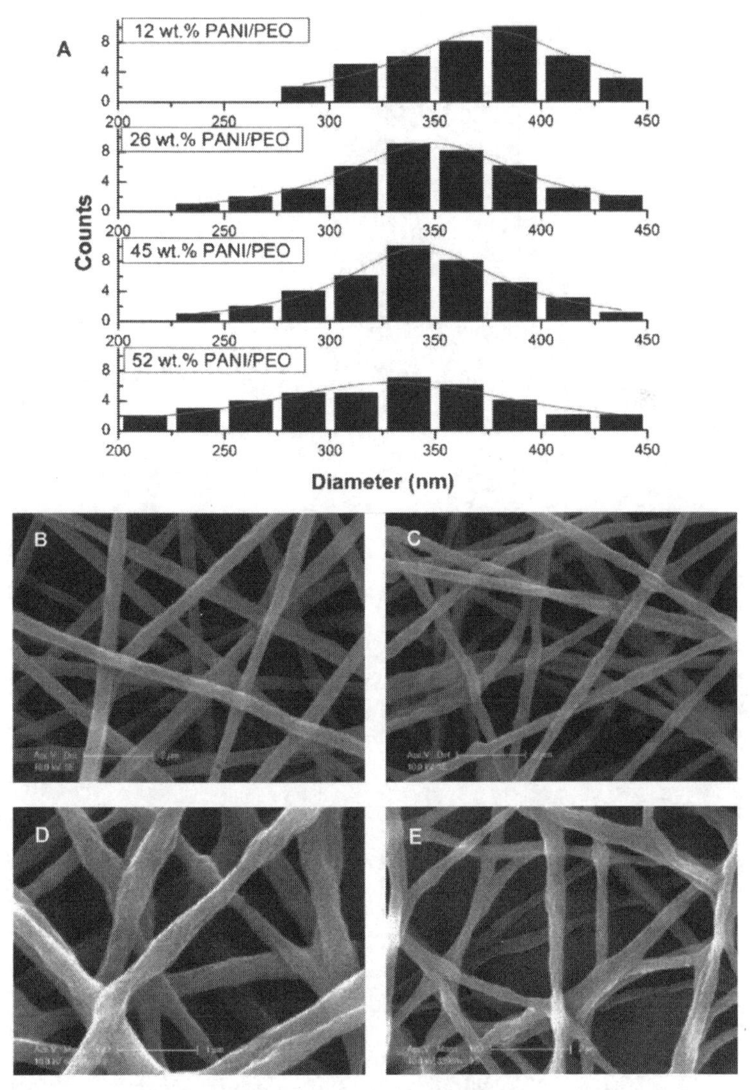

Figure 5.5 Characterization of fiber diameter and morphology of as electrospun PANI/PEO nanofibers. (A) fiber diameter distribution. Scanning electron microscopy images of nanofibers with PANI wt.%: (B) 12, (C) 26, (D) 45, (E) 52. Reprinted with permission from Ref. 155, Copyright © 2014 WILEY-VCH Verlag GmbH & Co. KGaA, Weinheim.

sensitivity toward NH_3 increased as the PANI content increased, but branched nanofibers reduced sensing response. The humidity sensitivity changed from positive to negative as the PANI content increased.[143]

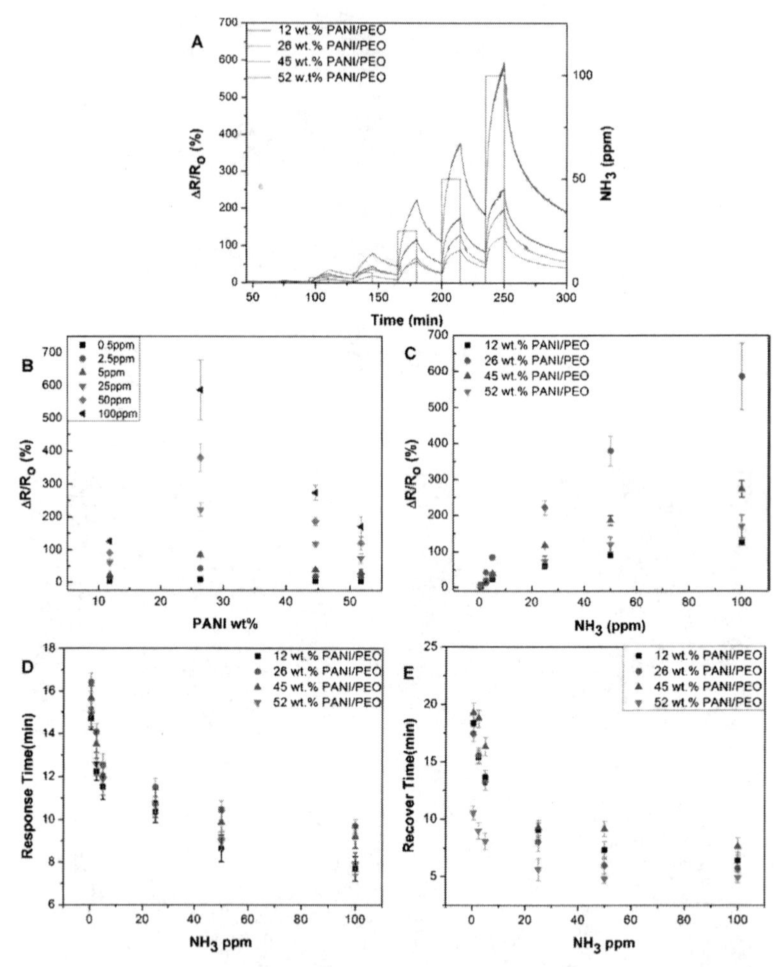

Figure 5.6 (A) Real time response of PANI/PEO composites with different compositions in response to various concentration of NH_3. (B–C) Normalized as fabricated sensor sensitivity. (D–E) Response and recovery time. Reprinted with permission from Ref. 155, Copyright © 2014 WILEY-VCH Verlag GmbH & Co. KGaA, Weinheim.

5.3.1 Effect of Rotating Speed on Fiber Alignment

By controlling the rotating speed of the disc, it was found that the fiber alignment differs by changing the rotating speed. The following scanning electron microscopy (SEM) image illustrate the effect of increasing the speed on the alignment of fibers, which is classified into five categories.

In the *steady state*, the electrospinning process was carried out while the rotating speed was zero, that is, not moving and steady. It was found out the fibers were random and formed as a web.

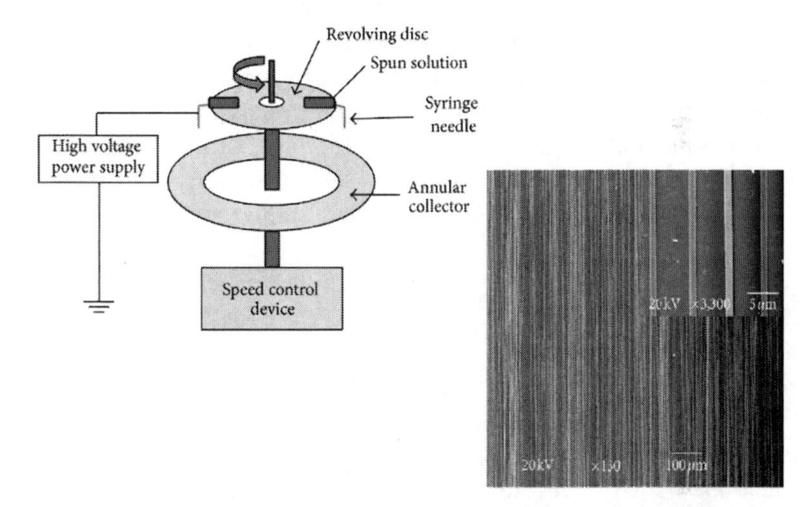

Figure 5.7 Schematic illustration of the centrifugal electrospinning apparatus (left). SEM images of aligned straight PS fibers electrospun at conditions of 3.0 kV, 420 rpm, and 2.5 cm. Inserted is a SEM image with larger magnification. Distributions of the degree of alignment defined by the difference between the angles of orientation of fibers and the long axis (right). Reprinted with permission from Ref. 156.

And at *low speed* by increasing the rotating speed of the disc to 10% (of 1500 rpm), the fibers were still in random arrangement with slight alignment. In centrifugal electrospinning, due to the electronic force partly taking the place of centrifugal force to get over the surface tension, the rotational velocity need not be very large. Experimental results indicate that the velocity from 300 to 600 rpm is large enough to meet the request. Faster velocity or slower velocity cannot facilitate the alignment degree of fibers. In order to simplify the experiments, the velocity-controlled device can provide three velocities, which let the disc revolve at three different rotational velocities (360, 420, and 540 rpm, respectively). Figure 5.7 shows the optical microscope image of PS fibers with 420 rpm rotational velocity. They were fabricated at the conditions that applied voltage was 6.2 kV and collecting distance was 3.0 cm. 420 rpm is found to be more suitable for the applied voltage of 6.2 kV, and under these conditions the polymer fibers fabricated are relatively parallel. Slower rotational velocity (360 rpm) cannot provide enough horizontal velocity to pull the jet fully straight.[156]

After adjusting the required speed to achieve good alignment of fibers, it is possible to collect those fibers into twisted yarn. Figure 5.8 describes the relationship between the velocity of jet and applied voltage qualitatively. In practical experiments, if the needle keeps static, the setup becomes conventional electrospinning; disordered fibers will be collected. With the increase of the rotational velocity, centrifugal force (rotational velocity) of the jet increases and partly replaces the electric force to overcome the surface tension, which leads to the decrease of applied voltage. Further, the horizontal motion of the needle can improve the fiber alignment. When the applied voltage decreases near to 0 V and the rotational velocity increases to thousands of rpm, the setup becomes centrifugal spinning. In this condition, it is difficult to produce highly aligned microfibers with uniform morphology. In fact, it has been observed that polymer fibers tend to be aligned when the perpendicular and linear velocities are very close or at the same order of magnitude. Thus, the optimum condition for aligned microfibers with fine morphology is limited in the black circle, as shown in Fig. 5.8.[156]

5.3.1.1 Yarn formation

After adjusting the required rotating speed, the electrospun fibers were collected and removed from the rotating disc. The electrospun fibers were twisted into yarn and had been tested by SEM to check the fiber arrangements. Figure 5.9 shows the twisted yarn and the direction of twist (a) and how the fibers prefer to align along the yarn axis (b). Better alignment of the fibers was obtained (Fig. 5.9) as the rotating speed of the disc was slightly increased.

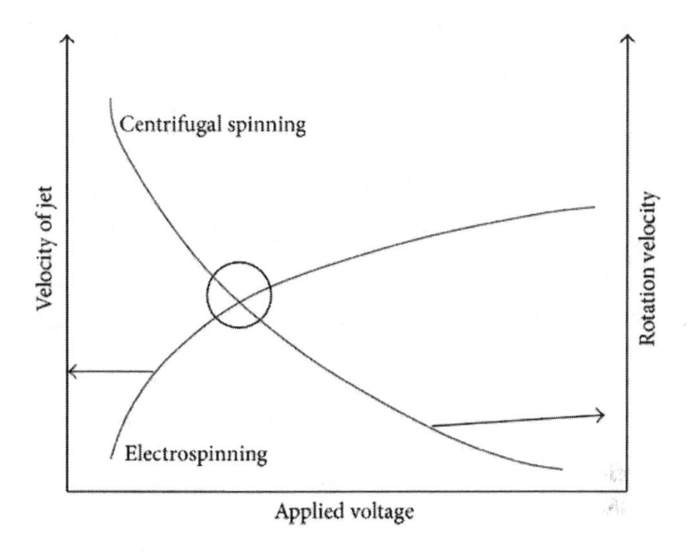

Figure 5.8 Schematic diagram of the relationship of applied voltage with velocity of polymer jet and rotational velocity. Reprinted with permission from Ref. 156.

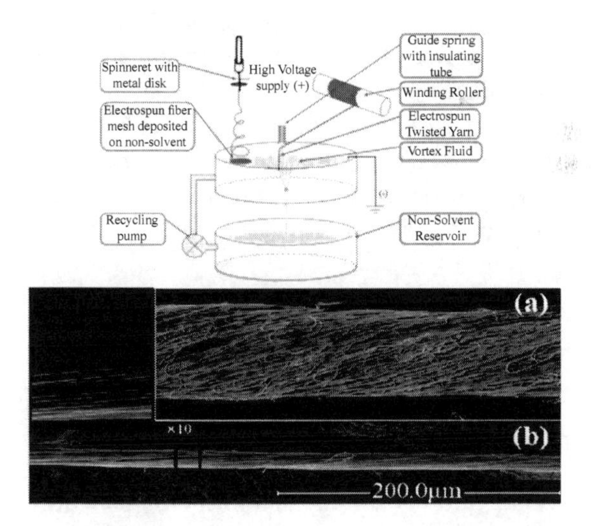

Figure 5.9 Nanofiber-yarn production schematic (left). Twisted nanofiber yarn at (a) 3 and (b) 4.5 m/min take-up velocities. Fiber alignment is clearly extensive (right). Reprinted with permission from Ref. 157, Copyright 2010, Society of Plastics Engineers.

5.3.2 Conductivity and Mechanical Properties of PEDOT Composite Fibers

Composite conductive fibers based on poly(3,4-ethylene-dioxythiophene)-polystyrene sulfonic acid (PEDOT-PSS) solution blended with polyacrylonitrile (PAN) were obtained via wet spinning. The influence of draw ratio on the morphology, structure, thermal degradation, electrical conductivity, and mechanical properties of the resulting fibers was investigated. The results revealed that the PEDOT-PSS/PAN composite conductive fibers' crystallization, electrical conductivity and mechanical properties were improved with the increase of draw ratio. The thermal stability of the fibers was almost independent of draw ratio, and only decreased slightly with draw ratio. Besides, when the draw ratio was 6, the conductivity of the PEDOT-PSS/PAN fibers was 5.0 S cm^{-1}, ten times the conductivity when the draw ratio was 2 (Fig 5.10).*

Poly(3,4-ethylenedioxythiophene) (PEDOT) nanofiber mats by electrospinning combined with in situ interfacial polymerization[159] (Fig 5.11). The PEDOT nanofiber mats displayed good mechanical properties (tensile strength: 8.7 ± 0.4 MPa; Young's modulus: 28.4 ± 3.3 MPa) and flexibility, which can almost be restored to its original shape even after serious twisting and crimping. Especially, from the results of the cellular morphology and proliferation of human cancer stem cells (hCSCs) cultured on the PEDOT nanofiber mats for 3 days, evidence was provided that the PEDOT nanofiber mats have similar biocompatibility to tissue culture plates (TCPs) combined with a high electrical conductivity of 7.8 ± 0.4 S cm^{-1}.

5.4 Characterization of PEDOT-PSS/PVAc Composites

5.4.1 FTIR-ATR Spectrophotometric Analysis

To improve the properties of nanofibers of poly(vinyl acetate) (PVAc), PEDOT-PSS was added to PVAc solution and composite nanofibers were obtained by electrospinning. The Fourier transform

*Excerpt reprinted with permission from Ref. 158, Copyright 2014, Royal Society of Chemistry.

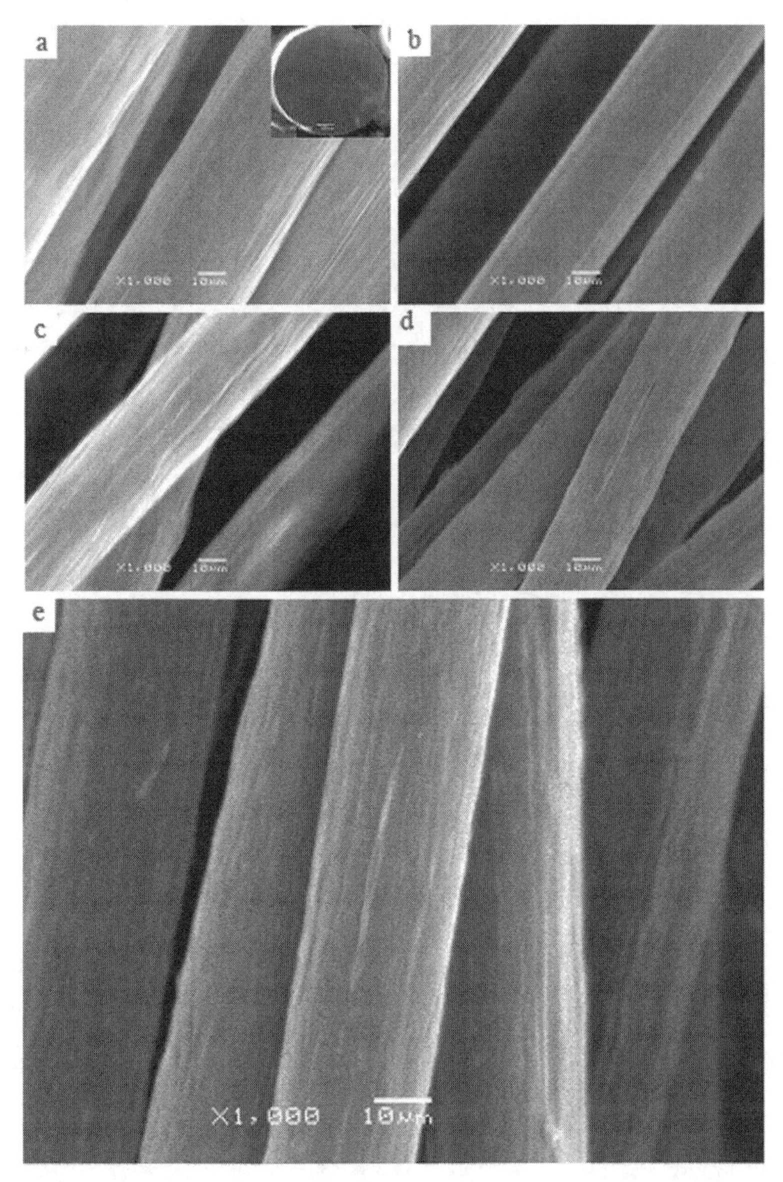

Figure 5.10 The SEM images of PEDOT-PSS/PAN composite conductive fibers with different draw ratios: (a) 2; (b) 3; (c) 4; (d) 5; (e) 6. Reprinted with permission from Ref. 158, Copyright © 2014, Royal Society of Chemistry.

infrared spectrophotometry–attenuated total reflectance (FTIR-ATR) spectra of PVAc and PEDOT-PSS/PVAc composites are presented in Fig. 5.12 and they were recorded in the absorbance mode.

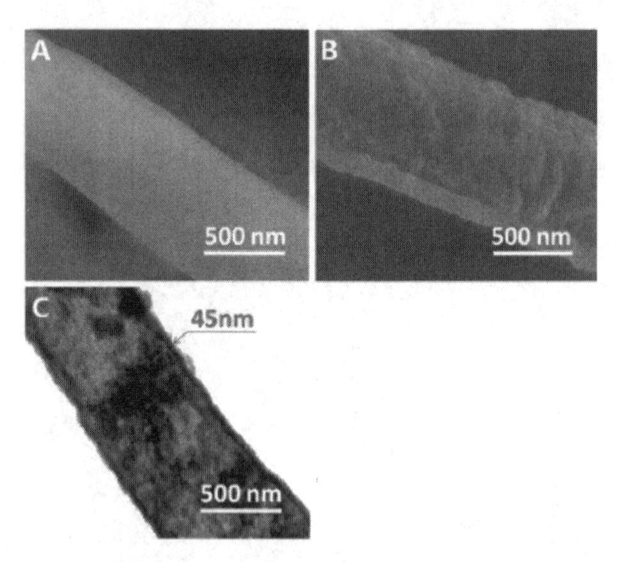

Figure 5.11 (A) SEM images of a PVC electrospun nanofiber with an average diameter of 610 nm, and (B) the obtained PEDOT nanofiber. (C) TEM image of a PEDOT nanotube. Reprinted with permission from Ref. 159, Copyright © 2013, Royal Society of Chemistry.

FTIR-ATR spectroscopic measurements were performed on PVAc and its nanocomposites. The spectra collected from nanofiber samples were almost identical to that of PVAc and not much changes observed.

Figure 5.12 shows the peaks at 1728 cm^{-1}, 1360 cm^{-1}, and 1224 cm^{-1} are assigned to C=O stretching of PVAc, C=C stretching vibrations of PVAc and PEDOT, and C–O stretching on the PVAc backbone and PEDOT, respectively. For the discrimination of C=C and C–O stretching bonds between PVAc and PEDOT; only a slight increment was seen due to their characteristic peaks being very close to each other. The peaks at 793 cm^{-1} were observed which probably belong to C–S stretching and C–H bending vibrations. Only slight differences can be found in their FTIR spectra, since composite solutions have a very small amount of PEDOT-PSS. PEDOT-PSS aqueous emulsion solution can be added to PVAc/DMF solution in slight amounts by preventing the precipitation. More than 1.5 g of PEDOT-PSS solution precipitates

PVAc in the composite solution. For this reason the ratio between PEDOT-PSS and PVAc was controlled to prevent precipitation.

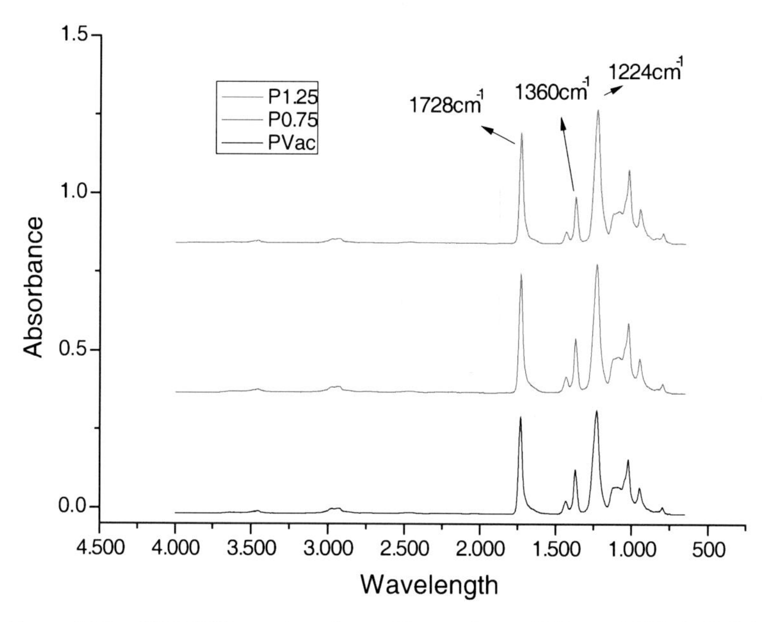

Figure 5.12 FTIR-ATR spectra of nanofibers of pure PVAc and PEDOT-PSS/PVAc with diferent amounts of PEDOT-PSS.[120]

5.4.2 UV-Vis Spectrophotometric Analysis

Nanofibers were characterized by UV-Vis spectroscopy. The composite nanofibers were dissolved in DMF, which were PVAc, P0.25, P0.50, P0.75, P1.00, P1.25, and P1.50 (details of compositions are given in Section 4.3.1).

Figure 5.13 shows the UV-Vis spectra of the solutions prepared from various PEDOT-PSS concentrations in PVAc. From this figure it can be seen that when compared to the spectrum of neat PVAc, regardless of the PEDOT-PSS concentrations in solution, all of the UV-Vis spectra (except for the PVAc spectra) exhibit a UV absorption in the range of 300–400 nm and an intense and broad absorption starting at larger wavelengths of ~700 nm. The high-energy transition is assigned to the n–π^* transition in the PEDOT backbone and the broader transition represents the free carrier tail that is common to conducting polymers (CPs).

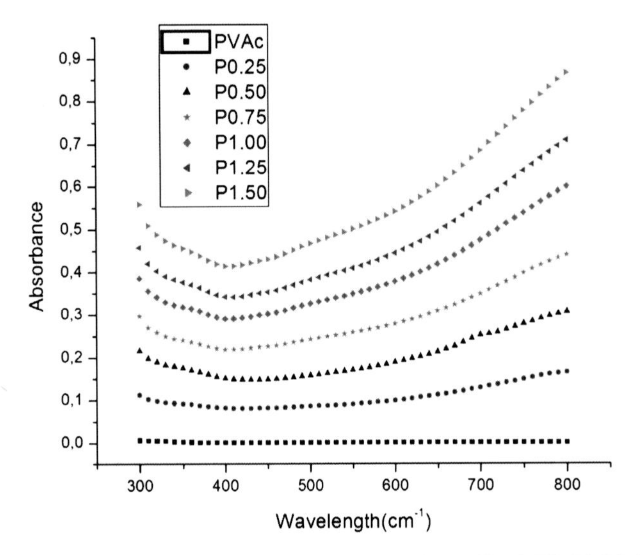

Figure 5.13 UV-Vis spectrum of PVAc and P0.25, P0.50 P0.75, P1.00, P1.25, and P1.50.[120]

As the amount of the PEDOT-PSS increases, absorbance also increases (Fig. 5.14). These results additionally imply that the PEDOT-PSS uniformly incorporated as a composite in the fiber.

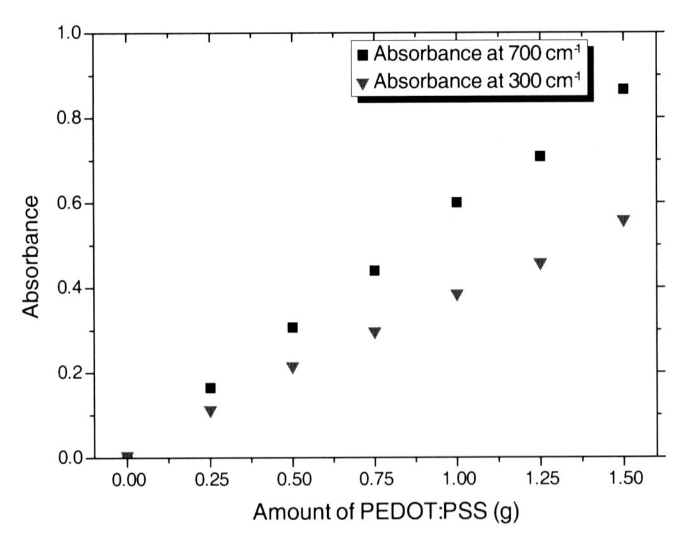

Figure 5.14 Relationship between the amount of PEDOT-PSS in a composite solution and the absorbances at 300 cm^{-1} and 700 cm^{-1}.[120]

5.4.3 Morphological Analysis

Morphological studies were performed by SEM observations of PEDOT-PSS/PVAc composites of electrospun nanofibers and the samples for the SEM measurements are coated with gold. The thickness of the gold coating was about 30 nm. Diameters of nanofibers were calculated using Image J software. SEM images show that morphology of the nanofiber mats of PEDOT-PSS/PVAc composites changed with different amounts of PEDOT-PSS. SEM pictures of electrospun nanofibers are presented in Fig. 5.15a–g.

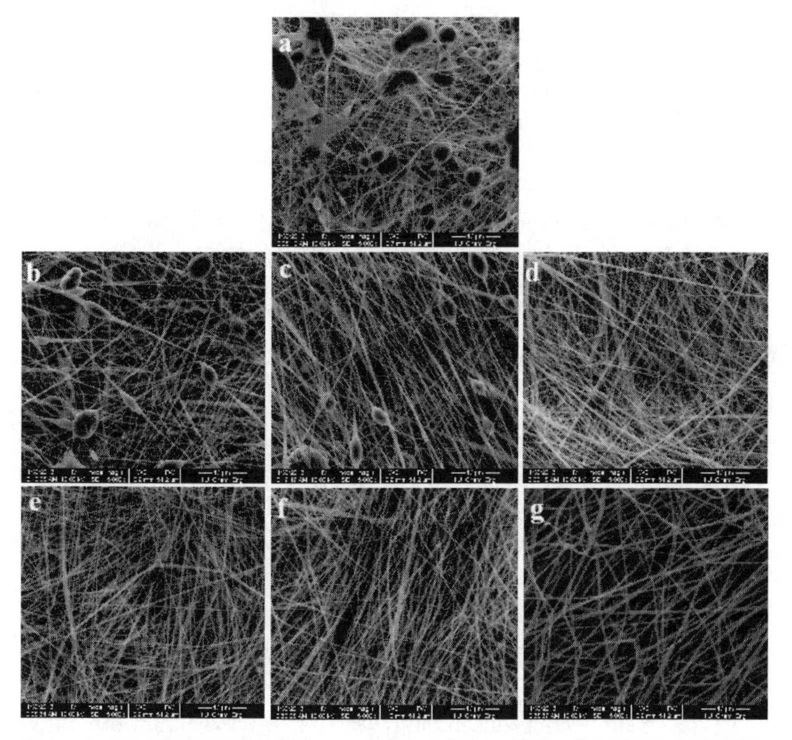

Figure 5.15 SEM images of the samples. (a) Neat PVAc, (b) P0.25, (c) P0.50, (d) P0.75, (e) P1.00, (f) P1.25, and (g) P1.50.[120]

According to the SEM images, the diameters of the PEDOT-PSS/PVAc nanofibers are dependent on initially added PEDOT-PSS concentrations in the solution mixture. At least 100 fiber diameters were measured for each sample. The diameter of PEDOT-PSS/PVAc fibers was larger than for pure PVAc fibers. The average nanofiber diameter increased from 139 ± 11 nm for pure PVAc fibers to 337 ± 8 nm for P1.50, as shown in Table 5.1.

Table 5.1 The effect of PEDOT-PSS content in solvent mixtures on fiber diameter

Sample Name	Avg. Diameter of NFBs (nm)
PVAc	139 ± 11
P0.25	139 ± 7
P0.50	151 ± 7
P0.75	188 ± 11
P1.00	194 ± 8
P1.25	215 ± 5
P1.50	337 ± 8

Source: Ref. 120.

An increased amount of PEDOT-PSS results in greater polymer chain entanglements within the solution, which is necessary to maintain the continuity of the jet during electrospinning. It is found that the PEDOT-PSS/PVAc solutions yielded bead-free nanofibers due to the greater polymer chain entanglements and viscosity of the solutions, but fibers were not observed for over 1.5 g PEDOT-PSS containing PEDOT-PSS/PVAc solution due to the high water content of PEDOT-PSS solution, which precipitates the polymer in DMF solution. So the amount of PEDOT-PSS is increased up to 1.50 g to obtain uniform bead-free PEDOT-PSS/PVAc nanofibers with a diameter of 337 ± 8 nm.

Significant changes in fiber diameter and morphology with different PEDOT-PSS contents was realized. By using different solvents and mixtures different fiber diameters have been observed. Figure 5.16 shows the relationship between PEDOT-PSS content and diameter of nanofibers where increasing the amount of PEDOT-PSS caused an increase in fiber diameter.

Experiments have shown that a minimum viscosity for polymer solution is required to yield fibers without beads, the solution of the P1.50 sample was determined the best electrospinning solution because of the continuous electrospinning process for this study.

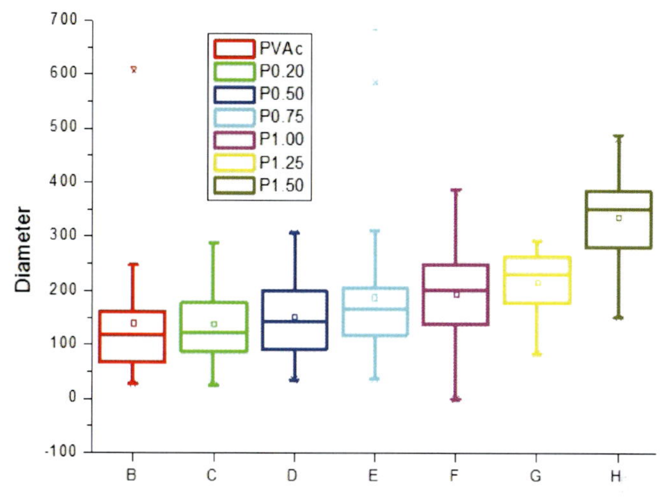

Figure 5.16 Relationship between diameter of nanofibers and amount of PEDOT-PSS.[120]

5.4.4 Dynamic Mechanical Analysis

Influence of the amount of PEDOT-PSS on mechanical properties was studied using dynamic mechanical analysis in tension mode. Uniaxial tensile tests were carried out at 30 ± 1°C with TA Q800 Dynamic Mechanic Analyzer. Elastic modulus and breaking points were measured by increasing ramp force 0.1 N/min to 18.0 N/min. Composite nanofibers were electrospun for three hours to measure mechanical properties. At least four specimens were tested for each measurement and the average values are presented.

Figure 5.17 shows stress–strain curves of neat PVAc and its nanocomposites in different PEDOT-PSS contents at 30 ± 1°C. The slope of the stress–strain curve in the elastic deformation region is the modulus of elasticity (Young's modulus—elastic modulus). It represents the stiffness of the material resistance to elastic strain.

PEDOT-PSS/PVAc composites showed brittle mechanical property. A decrease in tensile strength was observed for each composition in comparison to the PVAc and tensile strength decreased with increasing PEDOT-PSS content in the composition. However, afterward P1.00 sample yield decreases, leading to more brittle fractures due to a high value of PEDOT, which has a typical mechanical property of conductive polymers (brittleness).

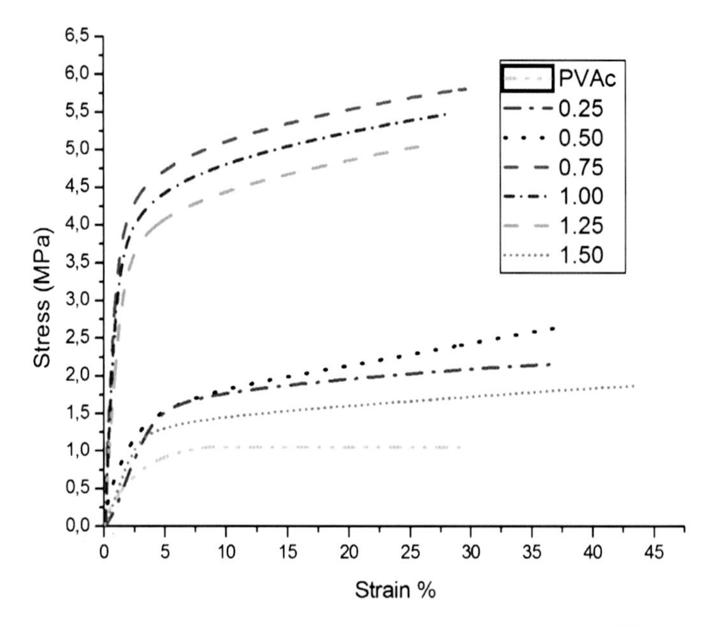

Figure 5.17 Stress–strain curves of neat PVAc and composites.[120]

The area under the stress–strain curve is toughness, which represents the total strain energy per unit volume in the material induced by the applied stress. It was seen that the toughness increases with increasing content of PEDOT-PSS in the composites up to P0.75. Increase in toughness is indicated in Table 5.2.

Table 5.2 Mechanical properties of PVAc and PEDOT-PSS/PVAc composites

Sample Name	Elastic Modulus (MPa)	Breaking Point (MPa)	Toughness (MPa)
PVAc	26.97	1.06	28.43
P0.25	35.43	1.53	66.74
P0.50	47.86	2.65	76.33
P0.75	384.57	6.51	151.57
P1.00	244.93	6.05	135.20
P1.25	182.06	5.31	113.06
P1.50	46.46	1.96	67.57

Source: Ref. 120.

Analyses show that sample P0.75 has the best mechanical properties compared to the other samples (Fig. 5.18). The modulus strength of PVAc is improved with the addition of PEDOT-PSS. However, the increment of modulus values decreases with a high amount of PEDOT-PSS, whereas the strength is increased with increased PEDOT-PSS content. It has been observed that the modulus of PEDOT-PSS/PVAc composites is higher than that of neat PVAc. It is also observed that the average value of the maximum modulus is reduced by increased PEDOT-PSS content, revealing the limited deformation and decreased ductility of PVAc.

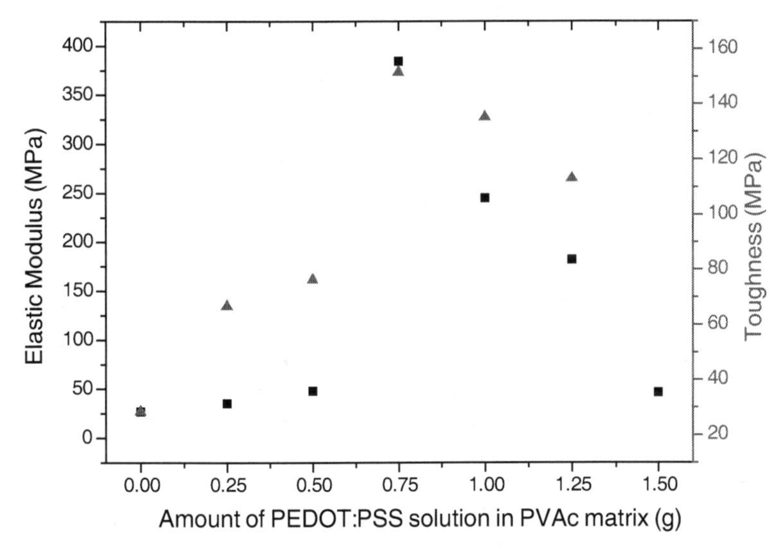

Figure 5.18 Relationship between the amount of PEDOT-PSS in composites and elastic modulus and toughness (square sign left, triangle sign right y-axes).[120]

5.4.5 BET Surface Area

ASAP2010M (Micromeritics, USA) is used—a fully automatic computer controlled setup for physical adsorption of gases (nitrogen, argon, carbon dioxide, etc) for pressures down to 1 mPa. The measurement is based on the adsorption and desorption of an inert gas on the whole surface of the sample at the same temperature. The adsorption and desorption isotherms are usually measured in the range of p/p_o 0.001 to 0.999 in liquid nitrogen. The surface area

(S_{BET}) is evaluated from obtained data by Brunauer–Emmet–Teller (BET) isotherm nonlinear parameter fitting.

Samples were measured by Krypton. Table 5.3 shows the results which is evident that the addition of even 0.5 g of PEDOT-PSS causes an increment of the surface area. However, no significant change was observed in the surface area with increasing content of PEDOT-PSS.

Table 5.3 Relationship between the amount of PEDOT-PSS and surface area (S_{BET})

Sample	Surface Area (S_{BET})
PVAc	3.6
P0.50	4.7
P1.00	4.6
P1.50	4.4

Source: Taken from Ref. 120.

5.4.6 Broadband Dielectric Spectrometer

Conductivity measurements were realized using broadband dielectric spectrometer. Nanofiber mat samples were prepared with the area of 4 cm^2 for measurement. As a result, conductivity values increased with an increase of PEDOT-PSS content at high frequency. Figure 5.19 shows the conductivity changes with the increase of the CP in the nanofiber composite.

PEDOT-PSS/poly(*N*-vinyl pyrrolidone) (PVP) and PVP nanofibers were independently prepared by electrospinning. The color of as-spun PEDOT-PSS/PVP nanofibers was light blue, indicating the incorporation of PEDOT-PSS in the nanofibers, whereas that of as-spun PVP nanofibers was white. The incorporation of PEDOT-PSS was confirmed from FT-Raman spectra.

Figure 5.20 shows the average diameter of PEDOT-PSS/PVP nanofibers as a function of applied voltage prepared at 9, 12, and 15 kV. The average diameter of PEDOT-PSS/PVP nanofibers increased as the applied voltage increased. When the applied voltage was 18, 21, and 24 kV, PEDOT-PSS/PVP nanofibers with corresponding diameters of ~154, 179, and 207 nm were fabricated. In general, higher applied voltage ejects more fluid in a jet, producing a larger fiber diameter.

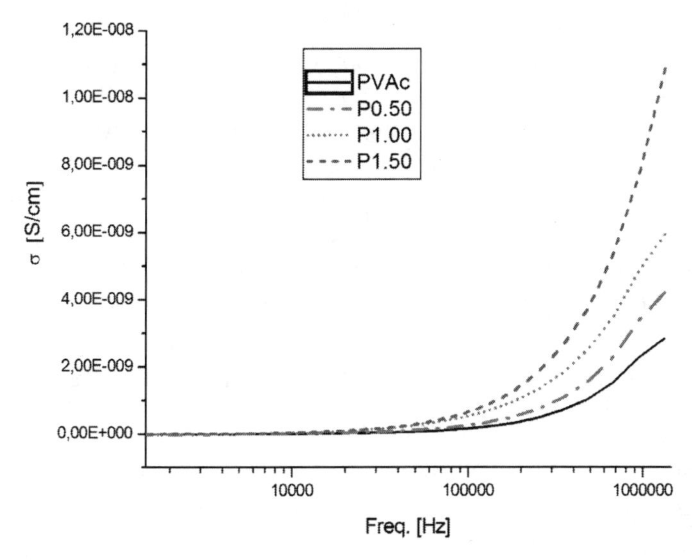

Figure 5.19 Conductivity curves of PVAc and PEDOT-PSS/PVAc composites.[120]

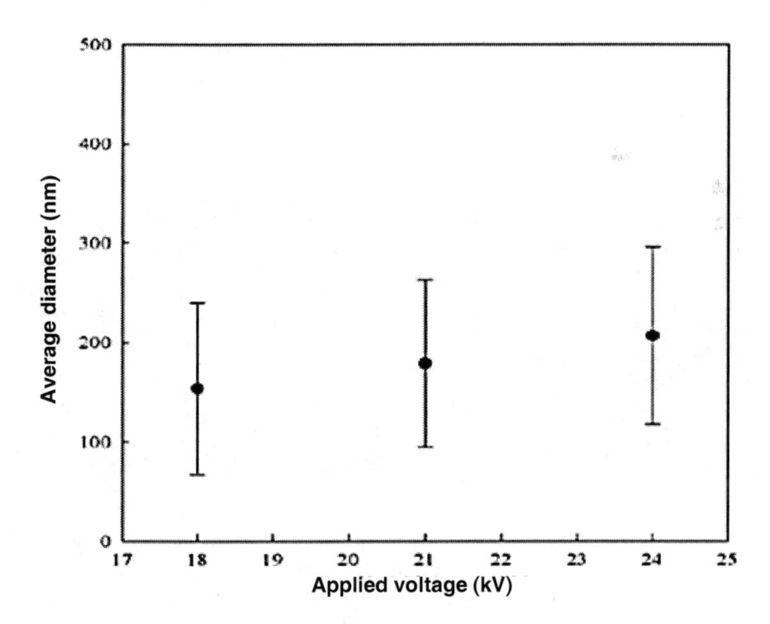

Figure 5.20 Average diameter of electrospun PEDOT-PSS/PVP nanofibers prepared from DMF as a function of applied potential (flow rate 0.2 mL/h; tip–collector distance 15 cm).[120]

The distribution of diameters of PEDOT-PSS/PVP nanofibers was insignificantly effected by the applied voltage. To compare PEDOT-PSS/PVP nanofibers with PVP nanofibers, electrospinning was performed under different conditions except the distance was constant between the tip and the collector (15 cm). PEDOT-PSS/PVP nanofibers were obtained while applying a voltage of 11 and 18 kV and a flow rate of 0.2 mL/h and the average diameters were found to be 153 and 212 nm, respectively. By incorporating PEDOT-PSS in PVP, the electrical conductivity of composite nanofibers was improved.

5.4.7 Organic Vapor-Sensing Characteristics of PEDOT-PSS/PVP and PVP Nanofibers

The sensing behaviors of PEDOT-PSS/PVP (Fig. 5.21) and PVP nanofibers on ethanol, methanol, THF, and acetone vapors are studied. The sensing was carried out for several cycles by repeated exposure of the nanofibers to saturated organic vapors and air alternately. Both PEDOT-PSS/PVP and PVP electrospun nanofiber sensors have exhibited good reversibility, reproducibility and response and recovery time. The response and recovery time of PEDOT-PSS/PVP nanofibers upon exposure to ethanol vapor are much faster than those of PVP nanofibers.

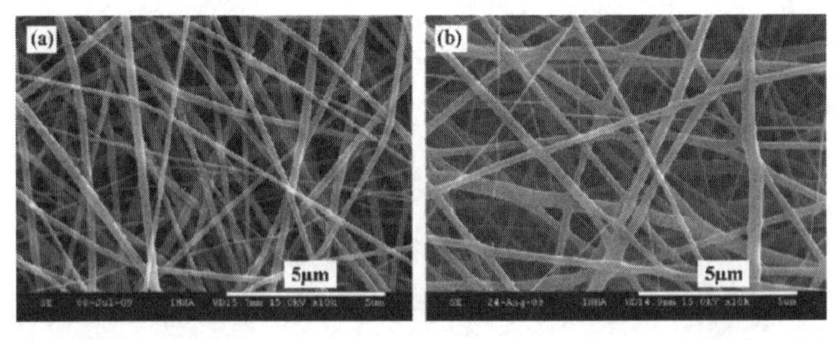

Figure 5.21 SEM images of electrospun nanofibers prepared from DMF solution. (a) PEDOT-PSS/PVP (applied voltage 18 kV; flow rate 0.2 mL/h; tip-to-collector distance 15 cm) and (b) PVP (applied voltage 11 kV; flow rate 0.1 mL/h; tip-to-collector distance 15 cm). Reprinted with permission from Ref. 160, Copyright © 2010 Elsevier Ltd.

By exposing PEDOT-PSS/PVP and PVP nanofibers to the solvents, these solvents produced opposite electrical responses in the PEDOT-PSS/PVP and PVP nanofibers. In the case of alcohol vapor sensing, the resistances of PEDOT-PSS/PVP and PVP nanofibers decreased and became constant upon certain saturation, but resistance was increased during alcohol vapor desorption by air. Such decrease may be associated with the dielectric constant of alcohols (Table 1.5).

Polar solvents with higher dielectric constants reduce the coulomb interaction between positively charged PEDOT and negatively charged PSS dopants by means of a screening effect between counterions and charge carriers.

Chapter 6

Impedance Spectroscopy and Spectroscopy on Polymeric Nanofibers

6.1 Impedance Spectroscopy on Polymeric Nanofibers

Electrochemical impedance spectroscopy is very useful technique to understand the microstructure and related further electrical properties of conjugated polymers. Optimization of properties of films is the key for a wide range of applications.[161]

A conjugated polymer modified electrode (CPME) can be onsidered as a heterogeneous system consisting of polymer film between the metal electrode and electrolyte. Two different interfaces behave as transition regions, and the properties are different from those of the bulk properties of the electrode, solution and the polymer film.

The impedance of a modified electrode depends on the impedance of each of the bulk phases and on interfacial properties as well. The equivalent circuit model (ECM) is used to elucidate the contribution of different charge transfer or transport processes to the overall impedance of electrodes. The equivalent circuit modeling concern finding equivalent electrical elements best representing physical processes within a range of frequencies by assuming each element

Nanofibers of Conjugated Polymers
A. Sezai Sarac

ISBN 978-981-4613-51-4 (Hardcover), 978-981-4613-52-1 (eBook)
www.panstanford.com

in the circuit represents a real process, the correct model assist to understand the effect of physical, chemical, and electronic properties on the charge transfer processes in conjugated polymeric electrode systems.[161]

6.2 Impedance Spectroscopy: A General Overview

In electrochemical impedance spectroscopy the cell is perturbed with a small time-varying potential wave-form $x = x_0 \sin(\omega t)$. The amplitude, x_0, is small, results in an output signal of the same frequency. This signal is defined by, $y = y_0 \sin(\omega t + \varphi)$, where φ is the phase angle. The ratio y/x is a complex number determining the impedance at the corresponding frequency. The real and imaginary components of impedance, Z_{real} and Z_{im}, and the phase angle depend on the particular character of the resistive or a capacitive behavior for certain frequency range. For instance, if resistive behavior dominates phase angle 0°, if the capacitive behavior dominates phase angle exhibits 90°.[161]

6.2.1 Equivalent Circuits Modeling

To model electrochemical behavior in connection with equivalent circuit composed of a number of single and/or sub-circuit elements. The capacitors and resistors can be obtained from the impedance spectrum (Fig. 6.1).

The complex impedance data involves the interplay of three variables, the imaginary component of the impedance, Z_{im}, the real component of the impedance, Z_{real}, and the phase angle, φ. To represent these variables as a function of frequency most common types of representation for impedance data are, the Nyquist and the Bode representations. Nevertheless, these have become the most widely used graphical representations of impedance data.

In the Nyquist format, the Z_{im} is plotted on the y axis and Z_{real} is plotted on the x axis (Fig. 6.1a–d). The impedance magnitude, $|\mathbf{Z}|$, for a circuit like the one shown in Fig. 6.1c is equal to the magnitude of a vector from the origin (0,0) to the point of interest (x,y) and the phase angle, φ, is given by the angle between the vector and the Z_{real} axis.[161]

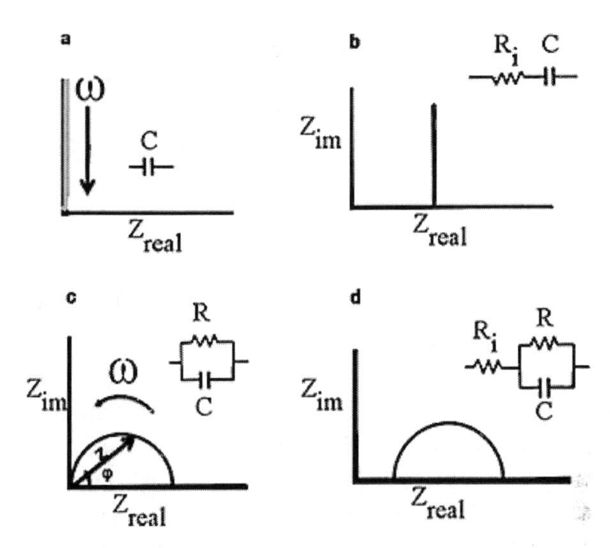

Figure 6.1 Nyquist plots for (a) a capacitor, (b) a capacitor in series with a resistor, (c) a capacitor in parallel with a resistor, and (d) a resistor in series with a parallel RC-circuit.

The phase angle can be accurately calculated from $\tan^{-1}\left(\frac{Z_{\text{im}}}{Z_{\text{real}}}\right)$.

The Bode format displays the phase angle *j* or the logarithm of impedance magnitude **Z** (*y*-axis) as a function of log frequency (*x*-axis). As such, the two types of plots yield complementary information.

The physical meaning of electrical elements used in the model should be supported by the comparison of simulated data and experimental results. To validate the physical meaning of electrical elements of model CPMEs, and to understand physical and electrochemical behaviour, the morphological structure of the polymer film (as rod-shaped fibrils, or polymer aggregates enclosing pores with poor interconnectivity between the aggregates) have been considered. The charge transport within the film and other macroscopic properties are dependent on the film texture and morphology.[161]

Other techniques and different geometric phase configurations are also required in order to understand the charge transport processes in CPMEs.

Generally experimentally proven models govern charge transport in terms of (1) charge transfer processes within conducting film, (2) charge transfer at substrate/polymer film interface, and (3) charge transfer at the solution/polymer film interface. Thus, an equivalent circuit can represent more physical meaning compared to graphical evaluations.

6.3 Modeling Charge Transport within the Conjugated Polymer Film

Conjugated polymers are different from metal conductors and inorganic semiconductors, and they have poor intrinsic conductivity. Conjugated polymers should be doped in order to conduct electricity, and their conductivity depends on doping level and preparation conditions.

Since inorganic semiconductors have crystallinity, charge carriers (electrons or holes) can be obtained by doping of inorganic semiconductors in the valence or conduction bands, respectively.[162]

On the other hand, doping of conjugated polymers can create accessible energy levels in the middle of the band gap, by maintaining valence and conduction bands remaining full and empty, respectively. In conjugated polymer films charge carriers do not move in a continuum carrier path but with intra-chain transport and charge recombination along the chains by the conductivity mechanism-hole/electron hopping. In this mechanism a series of redox processes takes place within the polymer film, causing a charge transfer resistance and a capacitance contribution to the impedance. In conjugated polymer film, the impedance behavior can be derived from several elements, the charge transfer resistance (R_{ct}), the resistance to electronic charge transport (R_e), and the capacitance related to charge separation (C).[161]

In addition to capacitors and resistors, equivalent circuit models include elements that do not have electrical analogs, i.e., as the Warburg (W) element and the constant phase element (CPE). These elements can explain the deviations from theoretical predictions of the models. The Warburg element is frequency-dependent, and its impedance may be represented by following equation:

$$Z_W = \frac{\sigma}{\sqrt{\omega}} - j\frac{\sigma}{\sqrt{\omega}} \tag{6.1}$$

where σ is the Warburg constant and ω is the angular frequency ($\omega = 2\pi f$). The Warburg constant, as it applies to CPs, can be obtained from the following equation:

$$\sigma = \frac{RT}{n^2 S^2 F^2 C_d \sqrt{2D}} \tag{6.2}$$

where F is Faraday's constant, D is the diffusion coefficient of diffusing species, n is the number of electrons transferred, S is the surface area of electrode, and R and T are the gas constant and temperature, respectively. The magnitude of the Warburg impedance depends on the frequency and on the rate of diffusion in the polymer film. A Warburg type impedance exhibits a straight line with 45° angle to the real axis in the Nyquist plot.

The Warburg behavior is attributed to the slow motion of ions within the film.[164] Including the driving force of ionic transport in the film, i.e., diffusion, migration or a combination of both.

In the case of $R_i = R_e$ diffusional behavior might be so small for certain conditions, and R_i is found to be a function of film thickness. Above a certain thickness, polymer film exhibits less homogeneity, leading to deviations from the ideal behavior.

The ionic diffusion rate is distributed across a range of values due to inhomogeneous composition in the direction perpendicular to the electrode surface where macro-inhomogeneous mode was considered.[165]

Homogeneous films exhibit non-linear diffusion which has been attributed to factors such as inhomogeneous thickness, charge trapping, or ionic charge adsorption in the film.[166]

In the inhomogeneous model where charge transfer takes place at interior pores filled with electrolyte can be expressed by using a frequency-dependent element CPE. Inclusion of CPE is generally associated with non-ideal capacitive behavior resulting from electrode roughness, inhomogeneous conductivity, or diffusion.[167] A CPE can be described by following equation:

$$Z_{CPE} = \frac{1}{A(j\omega)^n} \tag{6.3}$$

where A and n are constants. The constant n is related to the phase angle in the Nyquist plot, by the equation $\varphi = -(90n)°$. In CPE, n can take on values ranging between 0 and 1, showing the behavior of a resistive, capacitive, or diffusional component, as shown in Fig. 6.2. For example, for $n = 1$, the phase angle approaches $-90°$, the phase angle for a capacitor expresses its capacitance.

Transmission line circuits are used for homogeneous film with pores by considering the contribution of diffusion in the pores of the film.[168]

Transmission line models can be used for inert electrodes and it is a modification of the Randles model (Fig. 6.3). Since the Randles-circuit can be used to describe a nondistributed system, the transmission line models invokes a finite diffusional Warburg impedance, Z_w, in place of concentration hindered impedance (Fig. 6.4). Randles model is concerned with C_{dl} (the double layer capacitance), R_{ct} (the resistance to charge transfer) and Z_w by describing the processes occurring in the film. The expression of total impedance, Z_{tot}, is given by following equation:[161]

$$Z_{tot} = \left[\left(\frac{1}{Z_W + R_{ct}} + iwC_{dl}\right)S\right]^{-1} \tag{6.4}$$

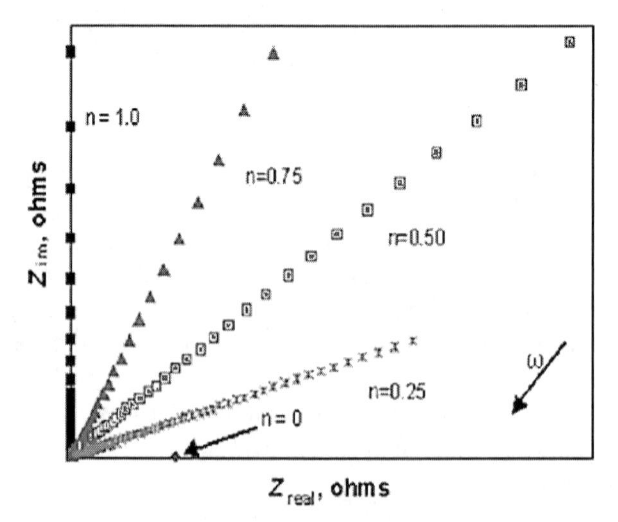

Figure 6.2 General behavior of a CPE as a function of n. Reproduced from Ref. 161 with permission of The Royal Society of Chemistry.

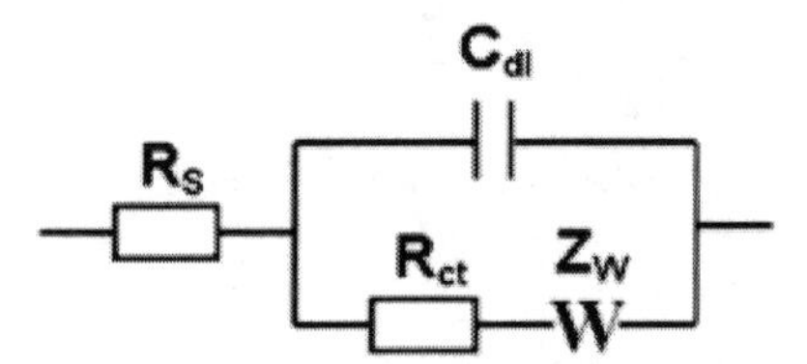

Figure 6.3 A Randles circuit where R_s is the solution resistance, C_{dl} is the double-layer capacitance, R_{ct} is the charge transfer, and Z_w is the diffusion-hindered impedance.

Randles model are used to describe the frequency dependence of diffusion and the capacitive impedance observed in the intermediate and low frequency ranges. A dual transmission line model has been proposed by including ionic and electronic resistance rails connected in parallel with a capacitance C_p (Fig. 6.5).[163] The model has been used to define the electrochemical behavior of polyaniline, and the capacitance was explained as a result of oxidation and reduction of the polymer. Ionic (R_i) and electronic (R_e) resistances are used to describe hindered motion of ions and electrons in the system, respectively. The impedance behavior has been found to be dependent on the ratio of the two resistances. [161]

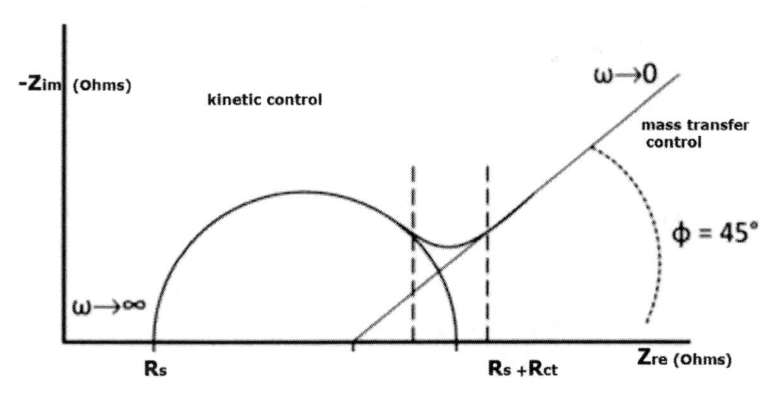

Figure 6.4 Impedance plane plot for a Randles equivalent circuit with charge transfer resistance and Warburg impedance. First region is a kinetics-governed semicircle tail. Last region is a mass transfer-capacitive tail. Region between two is a diffusion governed. Reproduced from Ref. 161 with permission of The Royal Society of Chemistry.

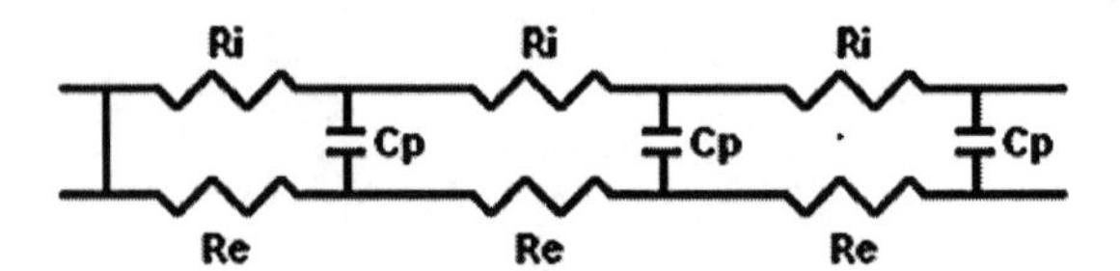

Figure 6.5 Transmission line circuit where R_e and R_i are the electronic and ionic resistances, while C_p is the redox capacitance. Reproduced from Ref. 161 with permission of The Royal Society of Chemistry.

6.4 Synthesis, Characterization, and Electrochemical Impedance Spectroscopy of PEDOT-PSS/PVAc and PEDOT/PVAc

The electrospinning method can be applied to produce nanofibers of PEDOT-PSS/PVAc and PEDOT/PVAc, which might be used for adhesives and coatings. For this purpose a commercially available aqueous solution of PEDOT-PSS was added in different amounts to the PVAc/*N,N*-dimethylformamide (DMF) solution, in addition to the polymerization of 3,4-ethylenedioxythiophene (EDOT) by cerium(IV) ammonium nitrate (CAN) in a PVAc matrix in DMF solution as well.

Products were characterized by Fourier transform infrared spectrophotometry–attenuated total reflectance (FTIR-ATR), ultraviolet visible (UV-Vis) spectrophotometry, scanning electron microscopy (SEM), and broadband dielectric/impedance spectroscopy (BDS). New absorption bands were observed corresponding to the conjugated polymeric units by FTIR-ATR and UV-Vis spectrophotometric analysis. The influence of concentration of PEDOT-PSS and PEDOT on the composite electrospun nanofibers was studied by EIS. Morphologies of electrospun nanofibers were also investigated by SEM.

Preparation details are the following: 0.1 g PVAc was dissolved in 10 mL DMF. Three series of polymer solutions were prepared with different amounts of 1.3 wt.% PEDOT-PSS aqueous solution, which were 0.25 g (P0.25), 0.50 g (P0.50), 0.75 g (P0.75), 1.00 g (P1.00), 1.25 g (P1.25), and 1.50 g (P1.50) each in 10 mL PVAc solution, and were stirrred at room temperature until homogenous solutions were obtained (Fig. 6.6a,b).

(a)

(b)

PSS

SO_3^- SO_3H SO_3H SO_3H SO_3^-

PEDOT

δ^-

H_3C

PVAc

(c)

Figure 6.6 (a) PEDOT/PVAc solutions: PVAc, S25, S50, and S75, respectively. (b) PEDOT-PSS/PVAc solutions: PVAc, P0.25, and P0.50, respectively. (c) Interaction of polymers.

For the synthesis of PEDOT in a PVAc matrix, 1 g PVAc was dissolved in 8 mL acetone and stirred magnetically at room temperature until a homogenous solution was obtained (Fig. 6.6c). EDOT was added dropwise to PVAc solutions; 25, 50, and 72 μL EDOT were used for the reaction (S25, S50, and S75, respectively). Synthesis was carried out with excess oxidant 1:2 (EDOT:ammonium cerium(IV) nitrate). A solution of CAN (respectively, 0.258, 0.515, and 0.773 mmol) dissolved in 2 mL acetone was then introduced at once.

The reaction mixtures were stirred for 24 hours at room temperature. The resulting dark-brown dispersions were used to prepare electrospinning solutions after adding DMF to the last solutions.

The distance between the tip of the syringe needle and the Al plate collector was 15 cm and the flow rate of the solution was 0.5 mL/h, while the electrical potential between the needle tip and the Al plate was 10 kV.

Obtained composite fibers were characterized by FTIR-ATR spectrophotometry, UV-Vis spectrophotometry, EIS, and morphological analysis.

The electrical properties of PEDOT-PSS/PVAc electrospun nanofibers on the indium tin oxide–poly(ethyleneterephthalate) (ITO-PET) surface were determined by electrochemical impedance measurements in a monomer-free solution 0.1 M. $NaClO_4$ in H_2O solution (Table 6.1, Fig. 6.7). The admittance plots of the solutions were in accordance with the conductivity increase (decrease in impedance) by the increase of PEDOT/PSS content.

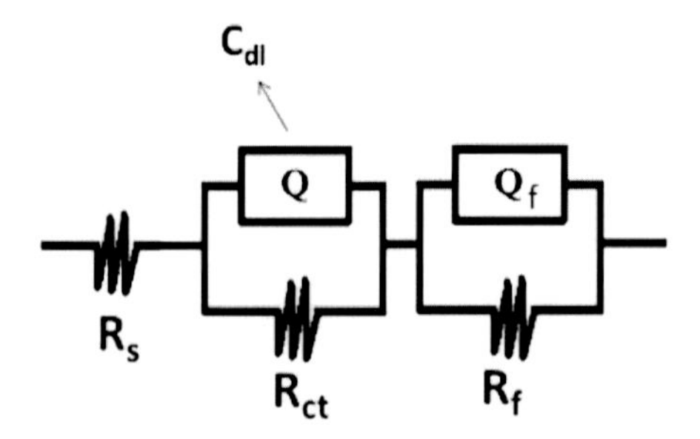

Figure 6.7 Equivalent circuit model of electrospun nanofibers of PEDOT-PSS/PVAc composites. R_s is the solution resistance, R_{ct} and Q (C_{dl}) are the resistance of the pore and electrolyte and the double-layer capacitance, and Q_f and R_f refer to capacitance of the nanofiber film on ITO-PET and resistance of the nanofiber film on ITO-PET, respectively.[120]

Table 6.1 C_{If} values for electrospun nanofibers of PEDOT-PSS/PVAc composites

	ITO-PET	P0.50	P1.00	P1.50
C_{If}(F)	4.24	20.00	10.00	10.08

Source: Taken from Ref. 120.

6.5 Characterization of Synthesized PEDOT in PVAc Matrix by FTIR-ATR, UV-Vis Spectrophotometric Analysis

To increase the amount of PEDOT in composite nanofibers, EDOT was polymerized in a PVAc/DMF matrix using CAN. The neat PVAc solution and resulting nanofibers were electrospun. The FTIR-ATR spectra of PVAc and PEDOT/PVAc composites are shown in Fig. 6.8 and it was recorded in absorbance mode.

FTIR-ATR measurement was performed on nanofiber mats of pure PVAc and its nanocomposites. The spectra were collected from all nanofiber samples that were compared to PVAc.

Figure 6.8 shows characteristic peaks of PEDOT at 740 cm^{-1}, 816 cm^{-1}, 1320 cm^{-1}, 1435 cm^{-1}, and 1646 cm^{-1}. Peaks at 740 cm^{-1} and 816 $cm^{-1}$are assigned to C–S stretching of the thiophene ring on PEDOT, and peaks at 1320 cm^{-1} and 1435 cm^{-1} are originated from symmetric and asymmetric C=C stretching vibrations of the thiophene ring, respectively. The peak at 1646 cm^{-1} corresponds to the C–C band of thiophene.

The solution of synthesized PEDOT containing PVAc and PEDOT-PSS/PVAc was followed by UV-Vis spectroscopy in DMF solution. The spectrum of the synthesized PEDOT/PVAc solution consists of bands at about 350, 450, and 550 nm. These peaks are characteristic for PEDOT absorption and confirm the polymerization of EDOT in the PVAc matrix (Fig. 6.9). A difference of maximum absorption of bands between PEDOT-PSS and synthesized PEDOT in a PVAc matrix was observed due to their oxidized and reduced forms.

SEM images indicated that the diameters of the PEDOT/PVAc nanofibers are dependent on initially added PEDOT concentrations in the solution mixture. The average nanofiber diameter decreased from 799 ± 66 nm for pure PVAc fibers to 409 ± 16 nm for S0.75, as shown in Table 6.2. Formation of PEDOT/PVAc nanofibers could not be performed as successfully as PEDOT-PSS/PVAc. It is related to the presence of oxidant in the polymer solution (Fig. 6.10a–d).

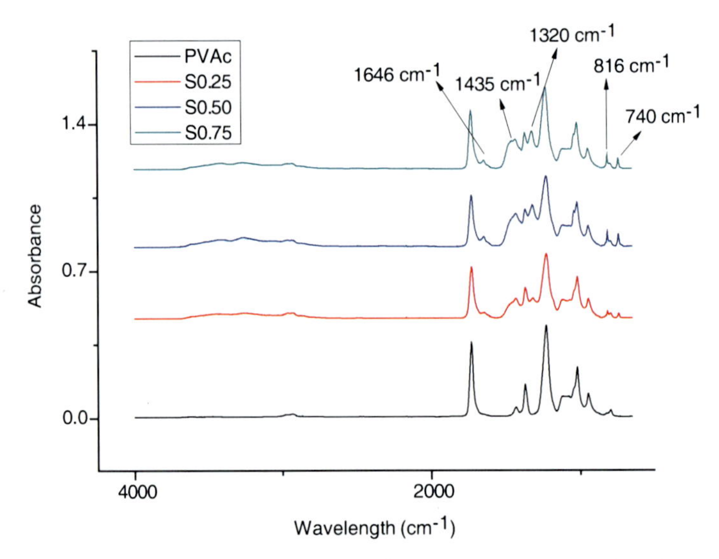

Figure 6.8 FTIR-ATR spectra of nanofibers of PVAc and PEDOT/PVAc (25, 50, and 75 μl EDOT: S25, S50, and S75, respectively).[120]

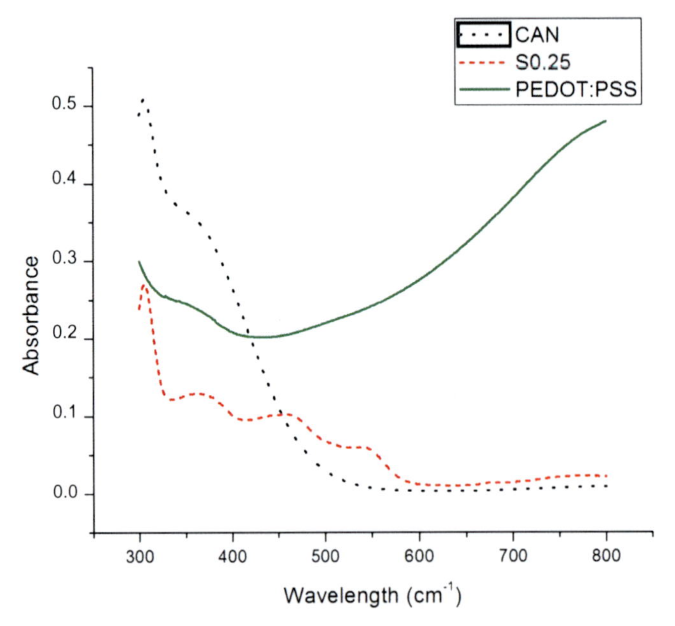

Figure 6.9 UV-Vis spectrum of the PEDOT-PSS, cerium(IV) ammonium nitrate, and PEDOT/PVAc composite electrospinning solution (with 25 μl EDOT).[120]

Table 6.2 Effect of PEDOT-PSS content on the fiber diameter

Sample Name	Avg. Diameter of NFBs (nm)
PVAc	799 ± 66
S0.25	1518 ± 24
S0.50	1936 ± 17
S0.75	409 ± 16

Source: Taken from Ref. 120.

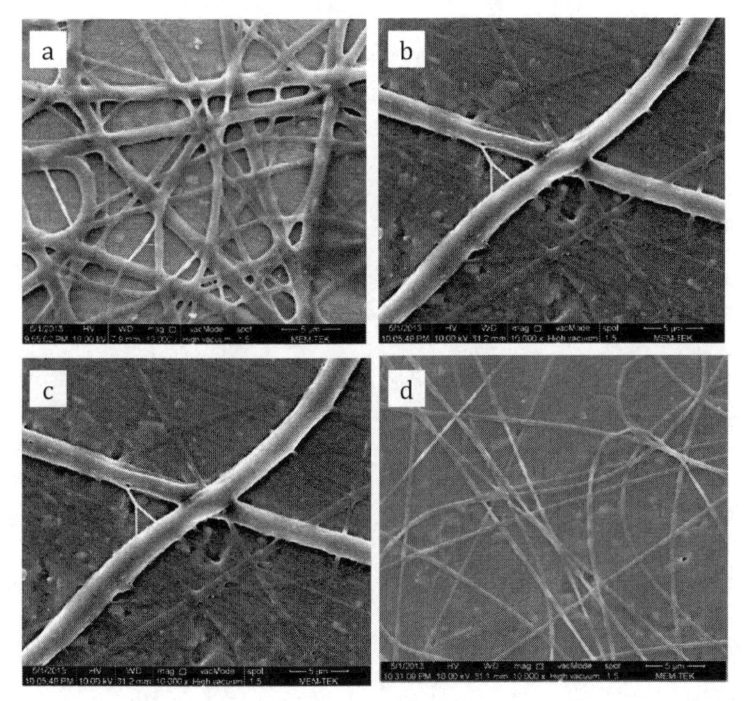

Figure 6.10 SEM images of the samples. (a) Pure PVAc, (b) S25, (c) S50, and (d) S75.[120]

6.6 Electrochemical Impedance Spectroscopy of Nanofiber Mats on ITO-PET

The electrical properties of these electrospun nanofibers were determined by electrochemical impedance measurements in a monomer-free solution (Figs. 6.11 and 6.12). $LiClO_4$ in 0.1 M in H_2O

solution was used as an electrolyte during impedance experiments. The samples were prepared by electrospinning of PEDOT/PVAc on the ITO-PET surface. Bode magnitude and admittance plots of the fibers were given in the frequency range of 0.01 Hz–100 kHz.

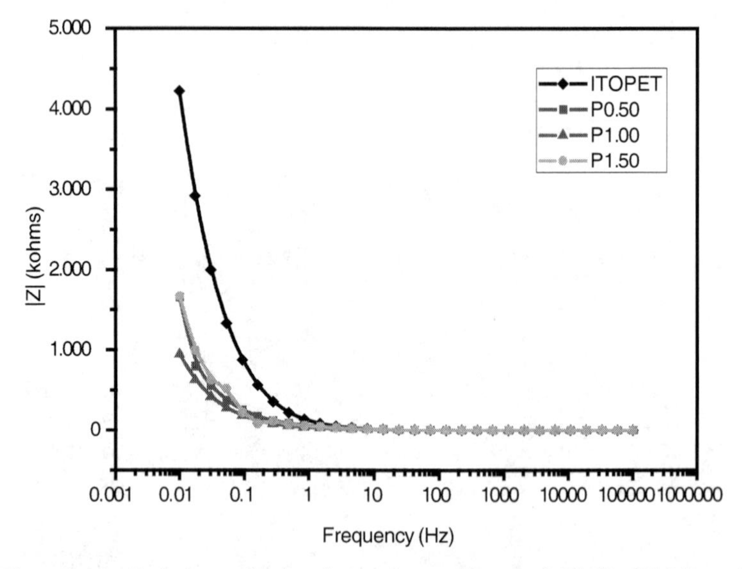

Figure 6.11 Bode magnitude plots of nanofibers of P0.50, P1.00, and P1.50.[120]

Low-frequency capacitances can be calculated from the following equation:

$$C_{\mathrm{lf}} = \frac{1}{2\pi f Z_{\mathrm{im}}} \tag{6.5}$$

where Z_{im} is obtained from the slope of a plot of the imaginary component of the impedance at low frequencies (f = 0.01 Hz). The C_{lf} of composite nanofibers decreased with the increase of PEDOT (Table 6.3).

Table 6.3 C_{lf} values for electrospun nanofibers of PEDOT/PVAc composites

	ITO-PET	S25	S50	S75
C_{lf}(F)	3.61	42.23	14.47	8.09

Source: From Ref. 120.

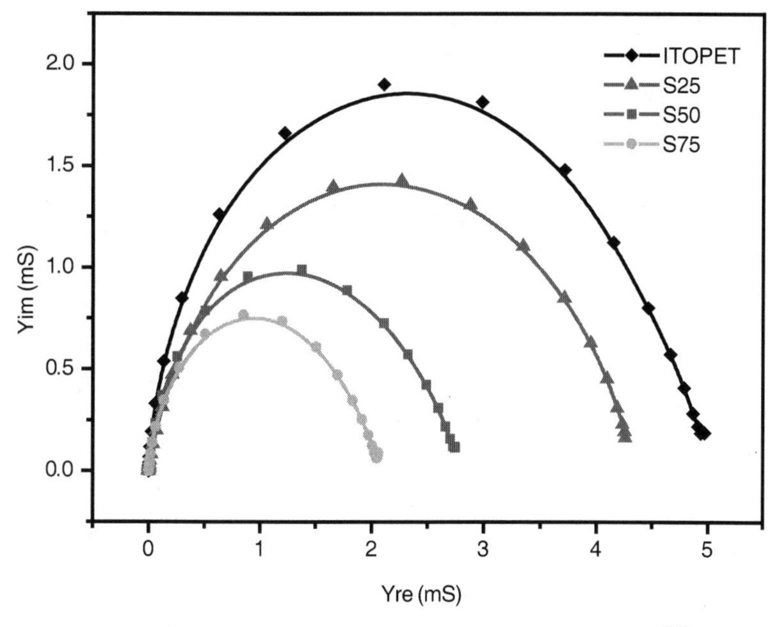

Figure 6.12 Admittance plots of nanofibers of S25, S50, and S75.[120]

The highest value of the semicircle admittance plot decreased with an increase of PEDOT in the composites due to the porous coating of semiconducting material and different morphology of fibers.

6.7 Molecular Structure and Processibility of Polyanilines

PANI is now accepted to have the general polymeric structure shown in Fig. 6.13. It differs from most other conducting electroactive polymers, such as polypyrroles (PPys) and polythiophenes, and it possesses three readily accessible oxidation states. These range from the fully reduced ($y = 1$) *leucoemeraldine* state to the half-oxidized (y = 0.5) *emeraldine* form to the fully oxidized ($y = 0$) *pernigraniline* state. PANI was the first sample of a doping conjugated polymer to a highly conducting regime by proton doping. Proton doping means that the *emeraldine base* (EB) form ($y = 0.5$) is doped with a protonic acid (e.g., 1.0 mol/L HCl) to produce a protonated EB form with high conductivity (~3 S/cm), which is called the *emeraldine salt* (ES) form. The proton doping does not involve a change of the number of electrons associated with the polymer backbone during the proton doping.

This significantly differs from redox doping (e.g., oxidation or reduction), which involves the partial addition (reduction) or removal (oxidation) of electrons to or from the polymer backbone. Thus proton doping is a major characteristic of PANI, differing from other CPs. The ES form is the state with the highest conductivity.[169]

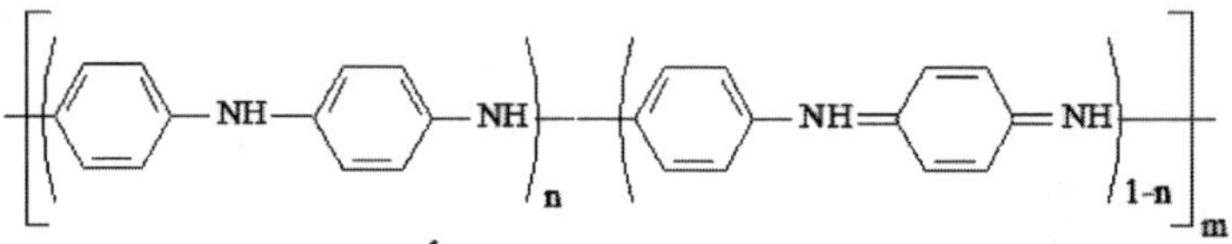

1

n = 1 (leucoemeraldine),0.5 (emaraldine) ,0.0(pernigraniline) / as basic forms

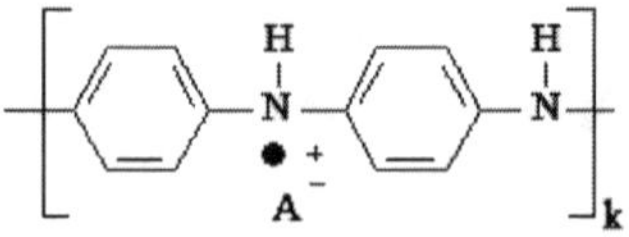

2

Emeraldine salt (ES) ,PANI.HA

Figure 6.13 Generalized composition of PANI indicating the reduced and oxidized repeat units: completely reduced polymer, half-oxidized polymer, fully oxidized polymer, and salt (ES) form, PANI, HA.

PANI differs from PPys and polythiophenes which N heteroatom participates directly in the polymerization process (PANI is a ladder polymer that polymerizes head to tail) and involves in the conjugation of the conducting form of the polymer to a greater extent than the N and S heteroatoms in PPy and polythiophene. PANI can be rapidly converted between base and salt forms by interaction with acid or base. These reversible redox and pH-switching properties, belongs to ES form (Fig. 6.14). Attractive fields for current and potential utilization of polyaniline is in, charge dissipation or electrostatic dispersive (ESD) coatings and blends, electromagnetic interference shielding (EMI), anticorrosive coatings, transparent conductors, actuators, chemical vapor and solution based sensors, electrochromic coatings (for color change windows, mirrors etc.), toxic metal recovery, non-volatile memory.

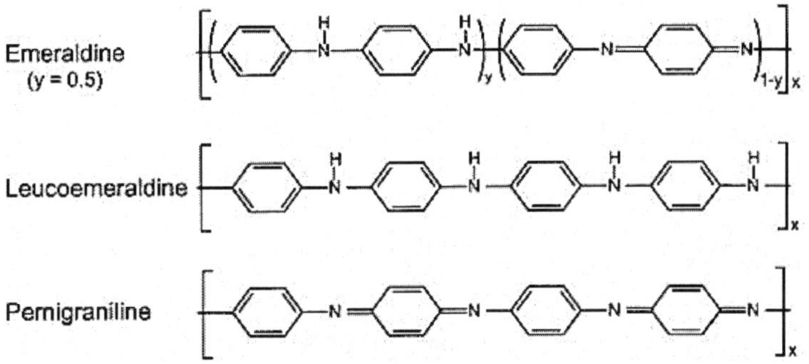

Emeraldine (y = 0.5)

Leucoemeraldine

Pernigraniline

Figure 6.14 Three major oxidation states of PANI: emeraldine (partially oxidized form), leucoemeraldine (fully reduced form), and pernigraniline (fully oxidized form).

6.8 Solubility and Processability

PANI has also interest as an organic magnet due to the potentially strong exchange interactions occurring through the conjugated backbone. However, for such applications, PANI must be processable, including being soluble in common solvents or melt process below the glass transition temperature. The EB form of PANI is soluble in *N*-methyl-2-pyrrolidone (NMP), which can be used to fabricate a freestanding film of the EB form. Its doped form (i.e., ES form) is insoluble in an organic solvent or aqueous solution. Therefore, synthesizing soluble conducting PANI (i.e., the protonated state, ES form) is a key to realizing application of PANI in technology. To solve solubility and processability of PANI therefore is necessary for not only commercial application but also fundamental research (i.e., structural characterizations). Efforts for improving solubility in solvents and processability of PANI have been reported. For instance, sulfonation or incorporation of *N*-alkyl-sulfonic acid pendant groups, dopant-inducted, self-doping polymers, microemulsion polymerization, and controlled relative molecular mass have been reported for improvement of solubility and processability of PANI.[170–173]

Correlations of solvent solubility parameters with molar attraction constants and with properties like surface tension, dipole moment, and index of refraction have been explored.[174]

It is possible to calculate the solubility parameters for polymers. A relation between the dispersion contribution to the surface energy of polymers (a measurable quantity) and the dispersion solubility parameter of polymers has been found, which is similar to a relation established for low–molecular weight substances (Table 6.4).

Table 6.4 Refraction index, dipole moment, and related solubility parameters for solvents

Solvent	Dispersion Solubility	Dipole Moment	Polar Solubility	Refractive Index
Methanol	7.42	1.70	6.0	1.3284
Ethanol, 99.9%	7.73	1.69	4.3	1.3614
n-Propanol	7.75	1.68	3.3	1.3855
n-Butanol	7.81	1.66	2.8	1.3993
Pentanol- 1	7.81	1.70	2.2	1.4100
Propylene glycol	8.24	2.25	4.6	1.4329
Ethylene glycol	8.25	2.28	5.4	1.4318
Cyclohexanol	8.50	1.86	2.0	1.4647
Ethyl lactate	7.80	2.40	3.7	1.4124
2-Butoxyethanol	7.76	2.08	3.1	1.4198
Oxitol (Cellosolve)	7.85	2.08	4.5	1.4077
Diacetone alcohol	7.65	3.24	4.0	1.4235
Diethylether	7.05	1.15	1.4	1.3524
Furan	8.43	0.71	0.9	1.4214
Dioxane	8.55	0.45	0.9	1.4224
Carbon disulfide	9.97	0.06	0	1.6279
Dimethylsulfoxide	9.00	3.90	8.0	1.4783
γ-Butyrolactone	9.26	4.12	8.1	1.4348
Acetone	7.58	2.69	5.1	1.3586
Acetophenone	8.55	2.69	4.2	1.5342
Tetrahydrofuran	8.22	1.75	2.8	1.4071
Ethyl acetate	7.44	1.88	2.6	1.3723

Solvent	Dispersion Solubility	Dipole Moment	Polar Solubility	Refractive Index
n-Butal acetate	7.67	1.84	1.8	1.3900
Isoamyl acetate	7.45	1.82	1.5	1.4007
Acetonitrile	7.50	3.44	8.8	1.3441
Butyronitrile	7.50	3.57	6.1	1.3838
Nitromethane	7.70	3.56	9.2	1.3811
2-Nitropropane	7.90	3.73	5.9	1.3943
Aniline	9.53	1.51	6.0	1.5862
Nitrobenzene	8.60	4.03	6.0	1.5500
Dimethylformamide	8.52	3.86	6.7	1.4304
Dipropylamine	7.50	1.03	0.7	1.4043
Diethylamine	7.30	1.11	1.1	1.3854
Morpholine	8.89	1.50	2.4	1.4542
Pyridine	9.25	2.37	4.3	1.5101
Carbon tetrachloride	8.65	0	0	1.4600
Chloroform	8.65	1.15	1.5	1.4460
Ethylene chloride	8.50	1.86	2.6	1.4448
Methylene chloride	8.52	1.14	3.1	1.4241
1,1,1-Trichloroethane	8.25	1.57	2.1	1.4379
Chlorobenzene	9.28	1.54	2.1	1.5248
o-Dichlorobenzene	9.35	2.27	3.1	1.5514
Benzene	8.95	0	0.5	1.5011
Toluene	8.82	0.31	0.7	1.4969
Xylene	8.65	0.45	0.5	1.4972

In the paint and coating industry, organic solvents may be classified by their solubility parameter. The solubility parameter is defined as the square root of the cohesive energy density and is expressed in units of MPa;[174] such units are often referred to as a Hildebrand (Ferdinand, R., 1989. *Principles of Polymer Systems*, 3rd ed., 28–37). The solubility parameters can be broken into three components representing nonpolar, polar, and hydrogen-bonding contributions (Table 6.5).

Table 6.5 Physical parameters of some solvents

Solvent	MW (g/mole)	BP (°C)	Evaporation Rate (Relative to *n*-Butyl-acetate)	Surface Tension (mN/m)	Hydrogen Solubility Parameter $(MPa)^{1/2}$	Polar Solubility Parameter $(MPa)^{1/2}$
H_2O	18.0	100	83	72	47.98	22.75
Acetone	58.08	56.5	1448	23.5	11.03	9.80
Methanol	34.04	64.5	590	22.07	24.00	13.01
THF	72.1	67.0	1227	26.4	6.67	10.97
MEK	72.1	79.5	631	24.6	9.47	9.25
DMF	73.09	152.0				
DMSO	78.13	189	0.026	43.53		
Ethanol	46.0	78.3	330	22.0	20.01	11.17
IPA	60.1	82.3	283	23.0	15.98	9.80
n-Propanol	60.1	97.2	130	23.7	17.68	10.54
Ethylene-glycol	62.07	197.6	0.4	48	29.79	15.08
Diethylene-glycol	106.1	245.8	0.01	44.7	23.32	12.28
Glycerol	92.06	290.1	0	64.0	31.41	15.41
1-Methoxy-2-propanol	90.1	120	N/A	27.7	27.83	14.73
n-Butanol	74.12	117.7	46	25.4	15.45	10.0
sec-Butanol	74.12	99.5	125	19.96	14.79	9.13
tert-Butanol	74.12	82.4	N/A	19.96	N/A	N/A
2-Methyl-1-propanol	74.12	108	82	23	14.96	9.80
Butyl Cellosolve	118.18	171.2	8	N/A	12.99	7.94

Source: Reproduced from Ref. 175.

6.9 Substitution onto the Backbone

Self-doped derivatives of PANI containing an ionizable, negatively charged functional group in their structure that acts as an inner dopant anion, bound to the polymer backbone. There is no anion exchange between the polymer and the solution during the oxidation or reduction process.

> These novel materials exhibit many properties that are different from those of the parent PANI and show promising applications in various fields. For instance, self-doped PANI is highly soluble in alkaline aqueous solutions and many nonaqueous media unlike PANI. Good solubility is essential for a polymer in order to facilitate postsynthetic processing. The solubility of PANI is greatly improved by the presence of the substituted acidic group.*

6.9.1 Sulfonated Polyaniline

The sulfonic acid group has been substituted with hydrogen on the benzene ring of PANI, resulting in a self-doped PANI.[177] By postsulfonation process of PANI resulting polymer becomes soluble in water. 3-Aminobenzenesulfonic acid was used in the synthesis of cytocompatible sulfonated polyanilines. It was used in fabrication of novel glucose biosensor having large active surface area and excellent conductivity.

It is also possible to polymerize the 3-aminobenzenesulfonic acid to produce polymetanilic acid which is also soluble. Thus, preparation of a solid polymer of this type is not possible in aqueous acidic solutions, but it may be possible in a neutral solution of aqueous–organic mixed medium. However, to exhibit the electrical and electrochemical properties, protonation of the imine nitrogen of the PANI backbone in poly(metanilic acid) is necessary, which requires an acidic solution. An acid group–substituted, self-doped PANI has better electrical and electrochemical properties over a wider pH range.

---

*Excerpt reprinted with permission from Ref. 176, Copyright 2010, John Wiley & Sons.

6.9.2 Poly(Anthranilic Acid)

Anthranilic acid (o-amino benzoic acid) is an important monomer for the synthesis of carboxylic acid group–substituted PANI (Fig. 6.15). Studies on the synthesis of poly(anthranilic acid) (PANA) from an aqueous acidic solution are scarcely reported in the literature, probably because of difficulty in synthesis, poor yield, and brittle nature of the film due to presence of an electron-withdrawing carboxylic group. PANA reveals high solubility in an aqueous solution of NaOH or NMP.[178] Similar to poly(metanilic acid), PANA exhibits electrochemical activity over a wide pH range in aqueous solutions owing to the substitution of the carboxylic acid group.

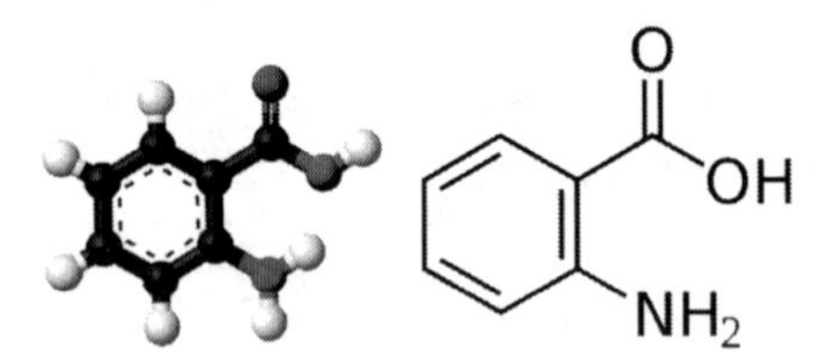

Figure 6.15 Structure of anthranilic acid, $C_6H_4(NH_2)COOH$.

PANA seems to be a promising material due to its high processibility, the presence of a redox-active substituent, and the ability to self-doping, all these characteristics being attributable to the carboxylic acid group. PANA is much more soluble than its parent polymer. PANA exhibits solubility in a range of solvents, including basic aqueous solutions,[179] alcohols, and other polar solvents.[180] The carboxylic acid group can be reduced and oxidized; this may provide an additional mode of charge storage (pseudocapacitance) for the supercapacitor. Self-doping has the advantage of being a faster doping process than the one seen in externally doped PANIs, where the dopant ion must move in and out of the polymer chain. The four oxidation states of PANA are presented in Fig. 6.16.[181]

In addition, self-doped PANI (Fig. 6.17) has an extended pH range of electrical conductivity and electrochemical activity, covering that of many biocatalysts and sensors.[182–185]

The limits of PANI applications in electrochemistry arise from the pH dependence of its conductivity. At pH > 4.0, PANI becomes electrochemically inactive. Incorporation of the acidic groups like carboxylic and sulfonic acids[186–188] as ring substituents influences the microenvironment of the amine groups and, namely, shifts the local pH, and then the conductivity does not fall off dramatically with increasing pH as in the case of PANI. Hence, the pH dependence of the electrochemical activity of the self-doped PANIs was improved markedly compared to the parent PANI that has a little electrochemical activity at pH > 4.0. Since carboxylic acid is an electron-withdrawing group and intramolecular proton exchange between acidic group and amine moiety, the oxidation potential of anthranilic acid is higher than the aniline, whereby the rate of oxidation and its polymer formation are largely lower than that of the aniline monomer.[189]

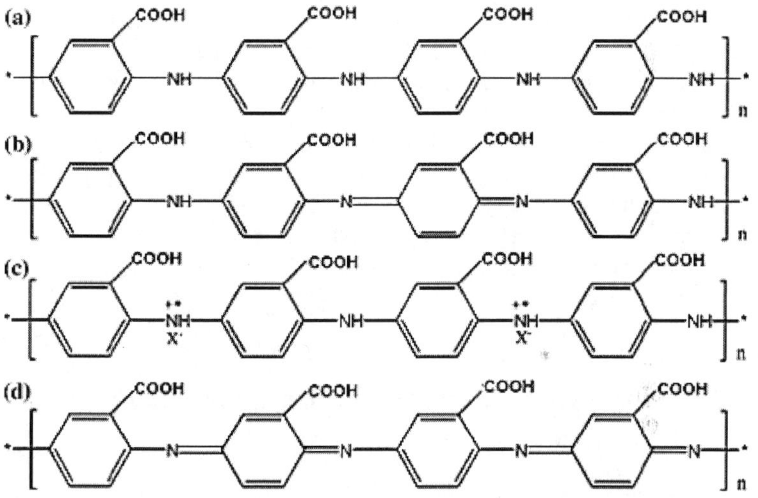

Figure 6.16 Four different redox forms of PANA: (a) leucoemeraldine base (fully reduced form), (b) emeraldine base (half-oxidized form), (c) conducting emeraldine salt (half-oxidized and protonated form), and (d) pernigraniline base (fully oxidized form).

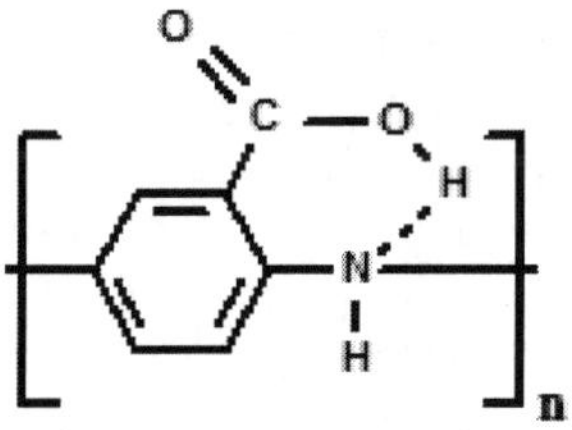

Figure 6.17 Proposed self-doped structure of the PANA.

6.9.3 Preparation of Poly(Anthranilic Acid)/Polyacrylonitrile Blends and Electrospinning Parameters

Different amounts of PANA were dissolved homogeneously in 10 mL DMF containing 0.5 g polyacrylonitrile (PAN) with enough viscosity for electrospinning. PAN solutions (without any PANA) were prepared and characterized to indicate the differences between PANA/PAN blends prepared by changing amounts of PANA (0.05 g, 0.075 g, and 0.1 g).[190]

> The electrospinning apparatus consisted of a syringe pump with a feeding rate from 5.5 μl/h to 400 mL/h, a high-voltage direct current (DC) power supplier generating a positive DC voltage up to 50 kV DC power supply, and a grounded collector loaded into a syringe. A positive electrode was clipped onto the syringe needle, having an outer diameter of 0.7 mm. The feeding rate of the polymer solution was controlled by a syringe pump and the solutions were electrospun horizontally on to the collector (Fig. 6.18). The polymer blends were electrospun to obtain nanofibers at room temperature at 15 kV driving voltages. The feeding rate was 1 mL/h and the distance between the capilary tip and the collector was constant, 15 cm.*

6.9.4 UV-Vis Spectrophotometric Investigation of Polymer Solutions

A composite solution containing different amounts of PANA was followed by UV-Vis spectroscopy in a DMF solution of nanofibers of PANA/PAN blends. The spectra consist of bands with the maxima located at about 536 and 370 nm. These peaks are characteristic for PANA absorption and confirm the presence of PANA in the nanofibers.[190]

PANA/PAN electrospun nanofibers, 0.1 g, with different initial mass ratios (including 0.05 g, 0.075 g, and 0.1 g PANA, respectively) were dissolved in DMF (20 mL) and peaks were exhibited between 340 and 380 nm, corresponding to π–π^* transition in the benzene ring of PANA. The broad absorption band extended from 480 nm to 650 nm due to the existence of polaron bands. These bands have been found to be dependent on the overall oxidation state of the

*Excerpt reprinted with permission from Ref. 190, Copyright 2013, Elsevier.

polymer. As the amount of the PANA increases, the hypochromic shift also rises (from 380 nm to 312 nm and from 650 nm to 480 nm). The hyphochromic shift illustrates that the carboxylate groups' steric effect in the polymer chain leads to perturbation of the coplanarity of the π system; thus, it decreases the level of conjugation and prevents charge transfer among the chains. Spectrum of PAN nanofibers dissolved in DMF (20 mL) was also investigated by UV-Vis spectrophotometry for comparison. Any peak was not observed from PAN nanofibers between 300 and 800 nm.

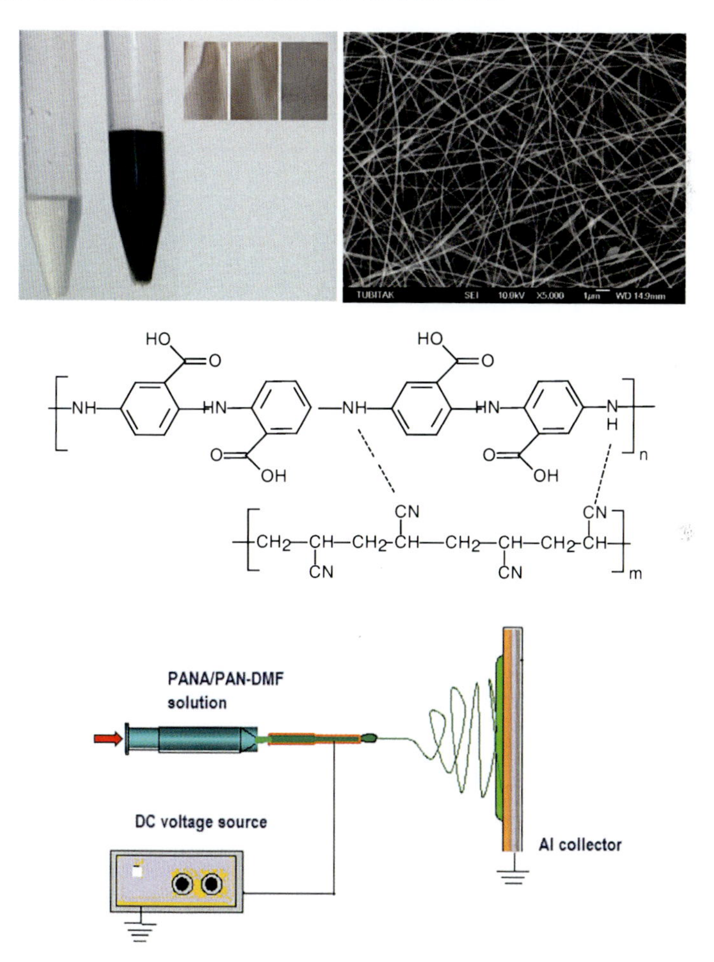

Figure 6.18 Electrospinning (DMF) solution with and without conjugated polymer (PAN, PANA/PAN), and nanofiber and electrospinning scheme. Reprinted with permission from Ref. 190, Copyright © 2013 Elsevier Ltd.

Several parameters, including chain length, the overall oxidation state of the polymer, interchain charge transfer, the extent of conjugation among adjacent phenyl rings, and the steric effect of the carboxylate groups in the polymer chain, indicate the real band positions of the spectra (Figs. 6.19 and 6.20) and the changes in these band positions are probably caused by these interactions.

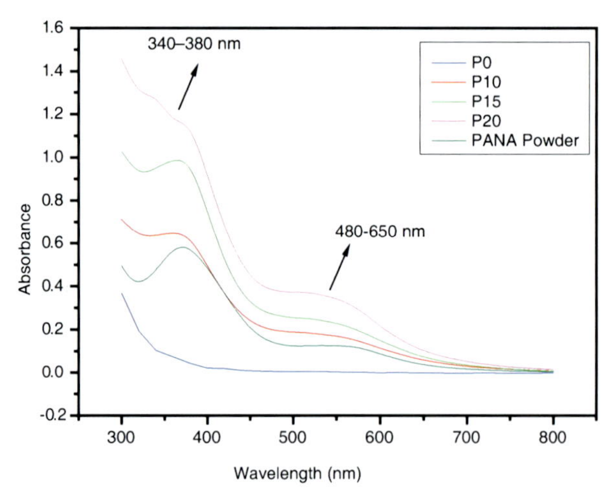

Figure 6.19 UV-Vis spectrum of the PANA/PAN nanofibers including different amounts of PANA. Reprinted with permission from Ref. 190, Copyright © 2013 Elsevier Ltd.

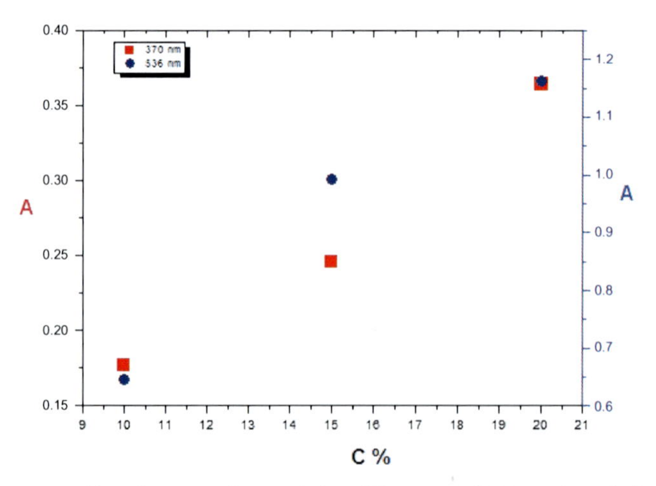

Figure 6.20 Absorbance values of the UV region (370 nm) and the Vis region (536 nm). Reprinted with permission from Ref. 190, Copyright © 2013 Elsevier Ltd.

Self-doping gives the benefit of being a faster doping process than the one illustrated in externally doped PANIs in which the dopant ion has to be in motion in and out of the polymer chain. The PANA/PAN interaction is presented in the Fig. 6.21.

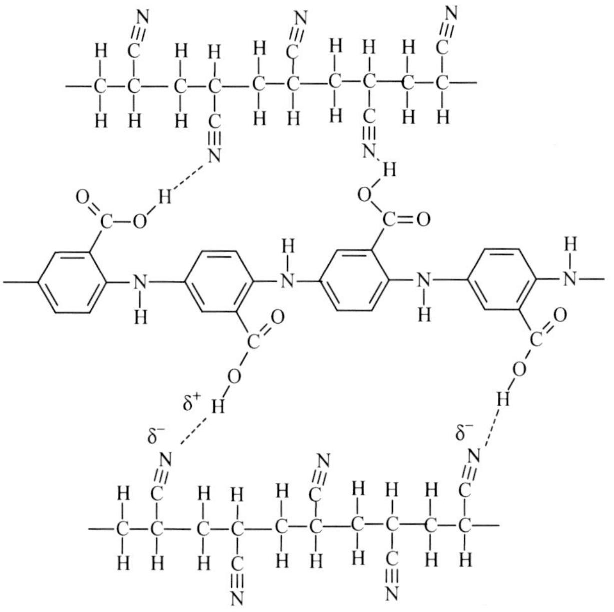

Figure 6.21 PANA/PAN interaction. Reprinted with permission from Ref. 190, Copyright © 2013 Elsevier Ltd.

6.9.5 FTIR-ATR Spectrophotometric Analysis of PAN/PANA Nanofibers

The FTIR-ATR spectra of PAN/PANA nanofibers are given in Fig. 6.22a,b, 3220 cm^{-1}, 1698 cm^{-1}, and 1514 cm^{-1} peaks, respectively, refers to as carboxylic acid–OH streching, C=O stretching, and C=C streching of benzenoid rings. In addition, the band appearing at 732 cm^{-1} probably corresponds to the C–H out-of-plane bending vibration of the 1-, 2-, 3-trisubstituted benzene rings. The peak at 1450 cm^{-1} refers to the C–O streching vibration of the carboxylic group proving that both –COOH and –COO groups are involved in the resulting polymer chain. These results provide an information about the presence of an emeraldine structure for PANA.

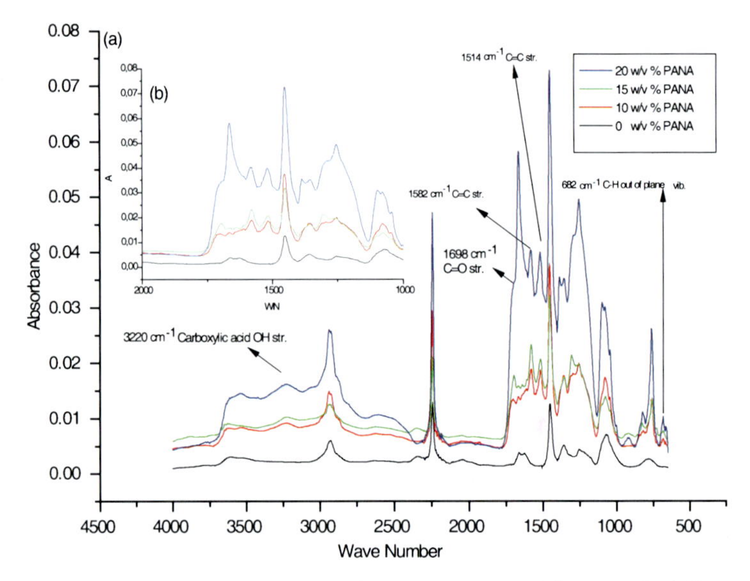

Figure 6.22 FTIR-ATR spectrum of PAN/PANA nanofibers with different amounts of PANA between 500 and 4000 cm^{-1} (a) and 1000 and 2000 cm^{-1} (b). Reprinted with permission from Ref. 190, Copyright © 2013 Elsevier Ltd.

Absorbance changes by FTIR-ATR spectra of PAN/PANA nanofibers with different contents of PANA and plots of absorbance versus corresponding peak ratios are displayed in Fig. 6.23.

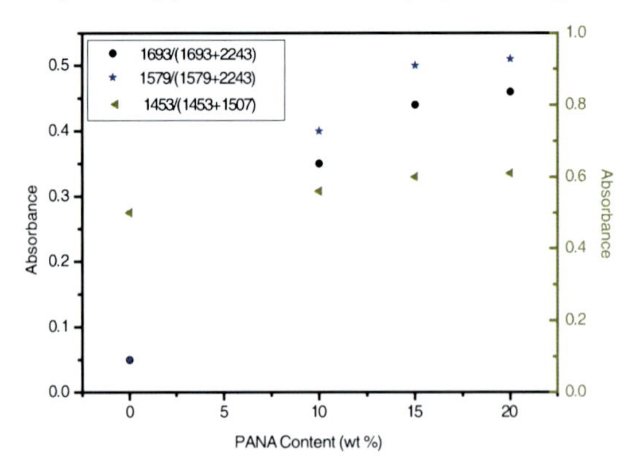

Figure 6.23 FTIR-ATR graph shows the increasing of absorbance values with the ratio of peaks vs. PANA concentration (w/v %), which also is refered to as the characteristic peak of DMF. Reprinted with permission from Ref. 190, Copyright © 2013 Elsevier Ltd.

6.9.6 Morphology of Electrospun Nanofibers of PANA/PAN Blends

SEM images show that morphology of the nanofibers of PANA/PAN blends changed with different amount of PANA of the mixed polymers. The morphologies of electrospun nanofibers are presented in Fig. 6.24a–d. According to the SEM images, the diameters of the PANA/PAN nanofibers are dependent on initially added PANA concentrations. The diameter of PANA/PAN fibers was smaller than for pure PAN fibers. The average nanofiber diameter decreased from 113 ± 23 nm for pure PAN fibers to 105 ± 23 nm for PANA/PAN (10 w/v %), 102 ± 20 nm for PANA/PAN (15 w/v %), and 94 ± 16 nm for PANA/PAN (20 w/v %), as shown in Table 6.6.

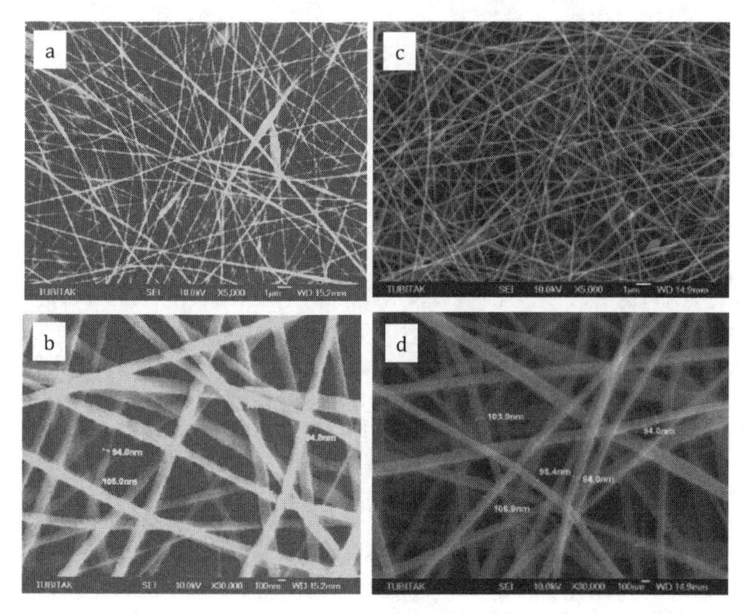

Figure 6.24 SEM images of nanofibers 0 w/v % PANA/PAN 5000x (a) and 30000x (b), and 20 w/v % PANA/PAN 5000x (c) and 30000x (d). Reprinted with permission from Ref. 190, Copyright © 2013 Elsevier Ltd.

PANA is homogenously distributed in the PANA/PAN blends. A similar trend was found for PANI and poly(lactic acid) (PLA). The diameter of the as-spun nanofibers was in the range of 500–800 nm, and that of the treated ones was in the range of 300–500 nm. From the SEM micrographs it is clearly seen that the homogeneous PANA dispersion in the PAN matrix is attained. This can be assigned to the interaction (hydrogen bonding) between two functional groups of PANA and PAN, namely, C–OH and CN groups and NH, which is the case of a low percolation threshold.

Table 6.6 Young's modulus values and average diameter of electrospun nanofibers (0, 10, 15, 20 PANA content w/v %)

Sample Name	Avg. Young's Modulus (MPa)	Avg. Diameter of NFBs (nm)
PANA/PAN 0 w/v %	20.56 ± 7.39	113 ± 23
PANA/PAN 10 w/v %	26.94 ± 7.89	105 ± 23
PANA/PAN 15 w/v %	23.91 ± 5.48	102 ± 20
PANA/PAN 20 w/v %	24.90 ± 2.14	94 ± 16

Source: Reproduced with permission from Ref. 190, Copyright © 2013 Elsevier Ltd.

6.9.7 Cyclic Voltammetry of Nanofibers of PANA/PAN Blends

The electrochemical behavior of nanofibers of PANA/PAN blends on an ITO-PET electrode was tested through cyclic voltammetric measurements in an acidic electrolyte including 1 M HCl and cyclic voltammograms (CVs) were recorded at a sweep rate of 20, 50, and 100 mV/s. Figure 6.25 depicts the oxidation/reduction process of PANA/PAN nanofibers on an ITO-PET electrode—an electrochemical cell consisting of an optically transparent ITO-coated PET (NV Innovative Sputtering Technology, Zulte, Belgium, PET 175 μm, coating ITO-60) as a working electrode and a Ag/AgCl and a Pt wire as a reference- and a counter electrode, respectively.

Unlike PANI and many of its substituted forms, PANA/PAN nanofibers do not show well-defined redox peaks in the potential range from –0.3 to 1.2 V. Among them, a pair of current peaks at about 0.2 V corresponding to the reversible transformation of leucomeraldine/emeraldine and another pair at about 0.7 V corresponding to emeraldine/pernigraniline reversible transition are prominent. The voltammogram is characterized by a pair of broad current peaks in the 0.5–0.6 V range, suggesting the electrochemical activity of PANA present in the PANA/PAN blends. The nature of this voltammogram differs from the voltammogram of PANI films because of the presence of the –COOH group on the PANI backbone.

In addition to this, different scanning rates were used to examine the electroactive property of PANA/PAN blends during the redox process (Fig. 6.25). The peak current was found to be proportional to the scan rates, suggesting an electric charge transfer–controlled process.

The formed nanofibers showed reversible behavior, which indicated formation of the cation radical and reduction of these sites in the electrolyte. This can be used as a sensing mechanism for certain ions. A high surface area and the presence of a carboxyl group result in a high current during oxidation and reduction. Figure 6.25 shows the stability of the film coating on ITO-PET with increasing number of cycles in a monomer-free solution.[190]

6.9.8 Dynamic Mechanical Analysis

Dynamic mechanical analysis of electrospun nanofibers of PANA/PAN were performed, the representative stress–strain curves of nanofibers are shown in Fig. 6.26. Their Young's modulus values were calculated and are shown in Table 6.6. As it was mentioned in the section on FTIR-ATR, the peak corresponding to the DMF at 1663 cm^{-1} may cause increasing of Young's modulus values as behaving a plasticizer.

6.9.9 Electrochemical Impedance Spectroscopy

The electrical properties of electrospun PANA/PAN blend nanofibers were determined by electrochemical impedance measurements in a monomer-free solution. The Nyquist, Bode magnitude, and Bode phase plots of the fibers were given in the frequency range of 0.01 Hz–100 kHz (Fig. 6.27a–c). The Bode phase angles, which approached the maximum at 15.20 Hz, were ~6.17° for 10% w/v, ~6.58° for 15% w/v, and ~7.27° for 20% w/v, respectively, Fig. 6.27c indicates the presence of PANA in the composite structure. The Bode phase peaks of the fibers in the frequency range of 10 000–100 000 Hz appeared, which should be caused by the transition from resistor to capacitor with the increase of the PANA ratio. The Bode magnitude plots for copolymers are presented in Fig. 6.27b, and PANA with 20% w/v had higher conductivity compared to 10% w/v and 15% w/v.

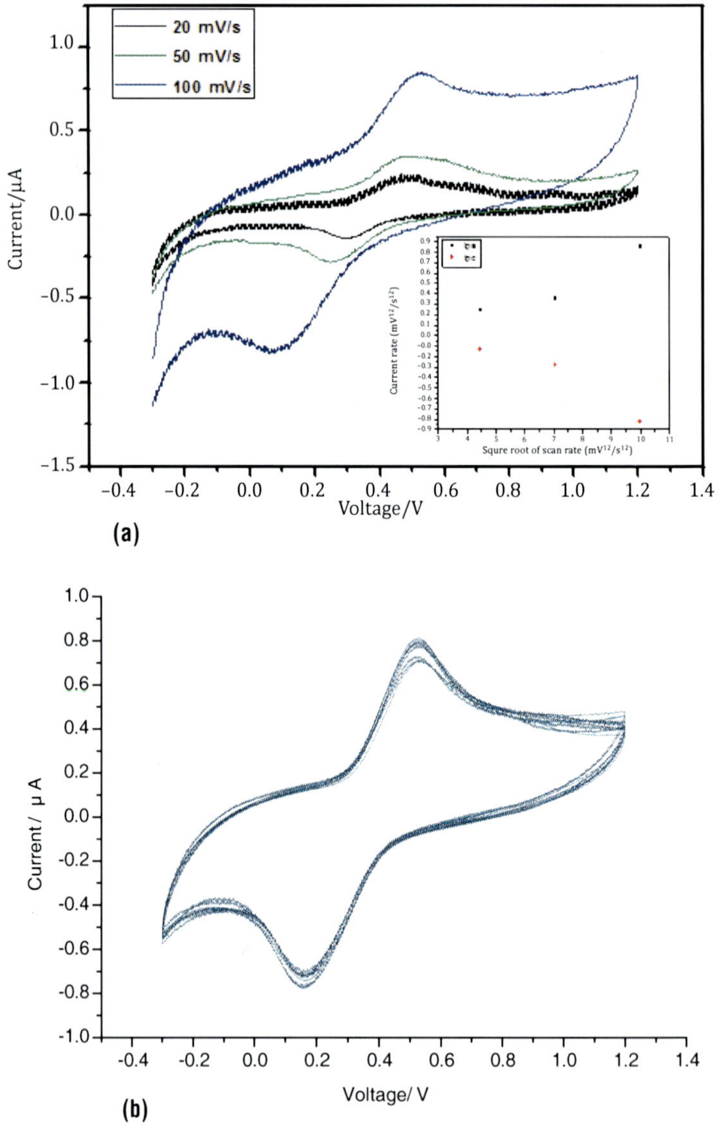

Figure 6.25 Cyclic voltammogram of nanofibers of 20 w/v % PANA/PAN with different scan rates as 20, 50, and 100 mV/s (a) and the inset graph indicates increasing of current values depending on scan rates. The stability of the film coating on ITO-PET with increasing number of cycles in a monomer-free solution (1 M HCl) (b). Reprinted with permission from Ref. 190, Copyright © 2013 Elsevier Ltd.

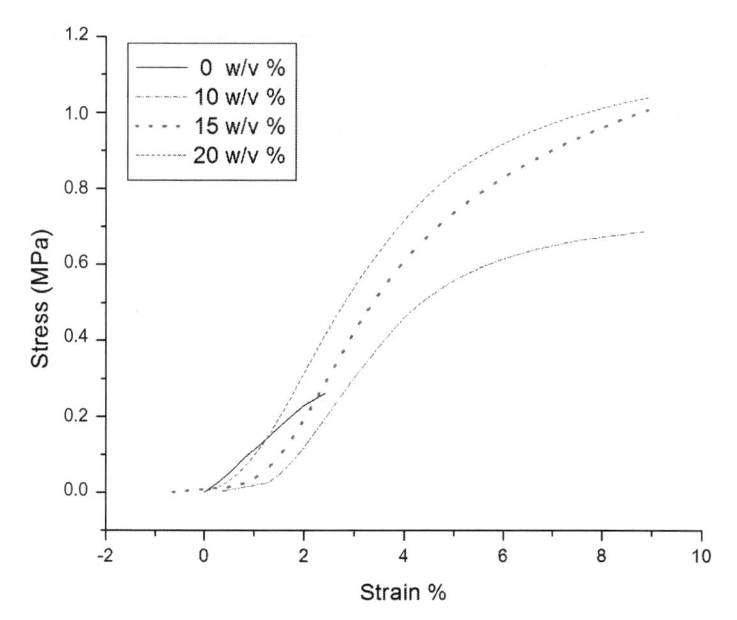

Figure 6.26 DMA analysis of nanofibers. Unpublished result of Digdem Giray and A. Sezai Sarac.

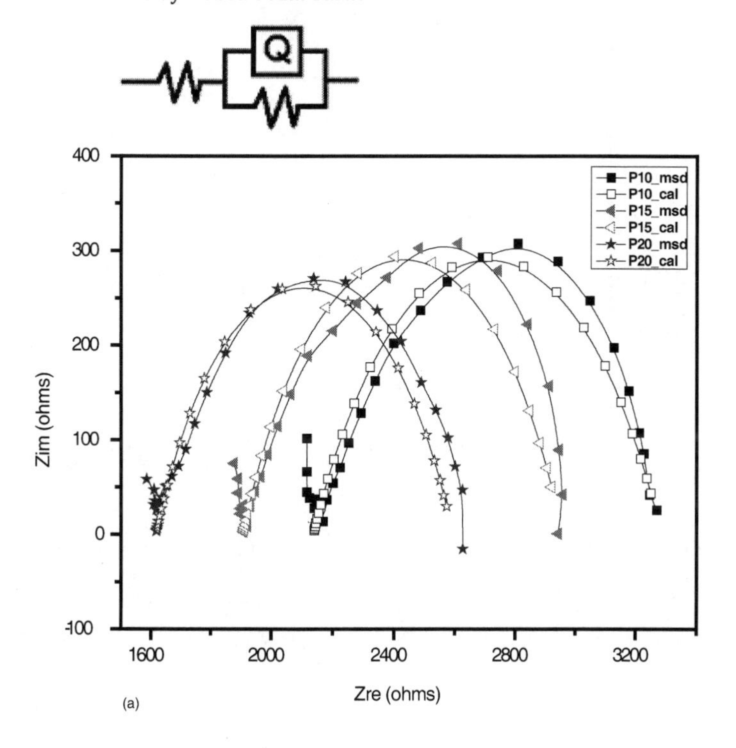

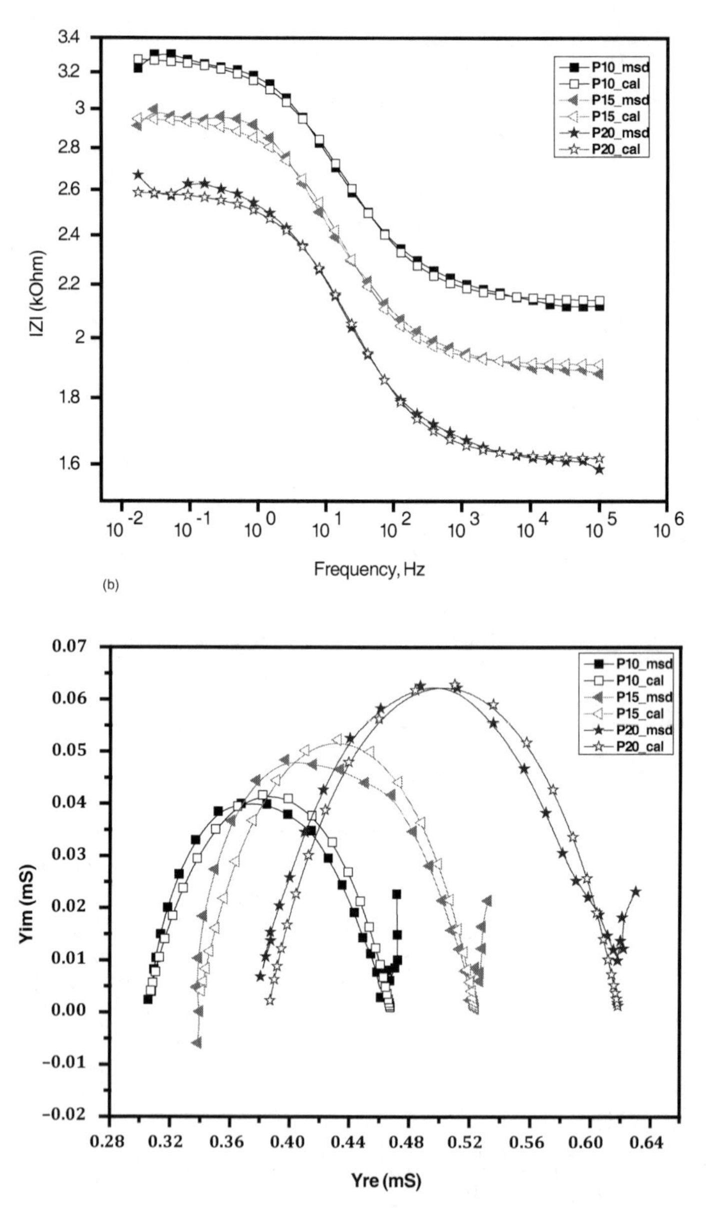

Figure 6.27 (a) Nyquist, (b) Bode magnitude, and (c) Admittance plots of different mole fractions of PANA/PAN composite nanofibers, correlated with the calculated data from the equivalent circuit modeling. (R(Q(R)). Reprinted with permission from Ref. 190, Copyright © 2013 Elsevier Ltd.

Nyquist plots were also used to estimate the low-frequency redox capacitance (C_{lf}) of the composite fibers. It can be calculated from the equation

$$C_{lf} = -\frac{1}{2\pi.f.Z_{im}} \quad (6.6)$$

where Z_{im} is obtained from the slope of a plot of the imaginary component of the impedance at low frequencies versus inverse of the reciprocal frequency f (f = 0.01 Hz) and other symbols have their usual meanings. Decrease in the charge transfer resistance (R_{ct}) of composite nanofibers was observed by the increase of PANA concentration in the blend (Table 6.7, Fig. 6.28). Due to semicircle behavior of Nyquist plots (complex plane plots) they can be a good candidate for biosensing measurements.[191]

Table 6.7 Content of PANA dependence of parameters calculated from the equivalent circuit model for the electrospun nanofibers of PANA/PAN blends

	10 w/v % PANA	15 w/v % PANA	20 w/v % PANA
R_S (Ω)	1385	1209	1230
CPE (Ω s^{-1})	5.42E-5	6.27E-5	7.42E-5
n	0.76	0.8	0.8
R_{ct} (Ω)	1167	1142	1041
Chi squared	1.47 e-03	1.04 e-03	1.09 e-03

Source: Reproduced with permission from Ref. 190, Copyright © 2013 Elsevier Ltd.

The diameter of the semicircle increased with an increase of PANA in the composite. It corresponds to the charge transfer resistance, which is included in the equivalent circuit as the resistance of the composite fiber.

The EIS data was fitted with an equivalent electrical circuit in order to characterize the electrochemical properties of the composite. The experimental results and values obtained from equivalent circuit modeling have shown that both the charge transfer resistance and the double-layer capacitance decreased with the increase of the incorporated PANA into the composite structure. The impedance spectra for nanofibers may be described by the equivalent circuit of (R(Q(R)).[190]

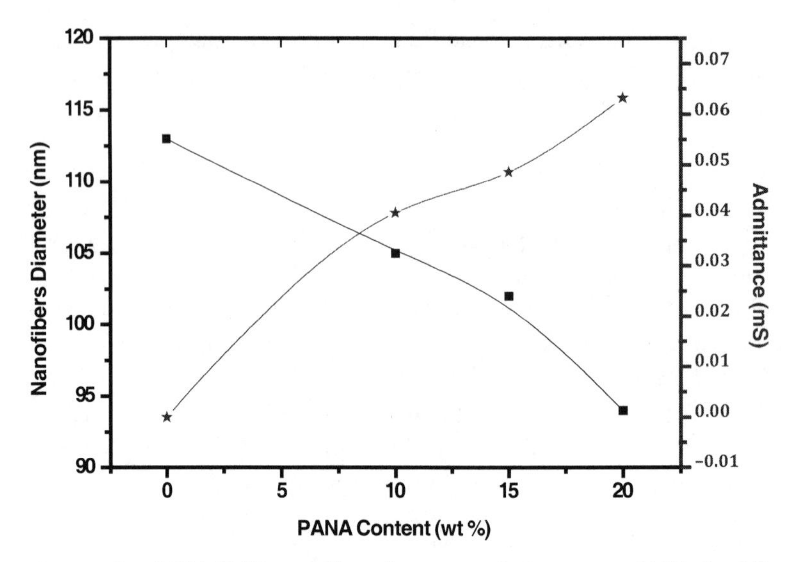

Figure 6.28 PANA/PAN nanofiber diameter; admittance vs. PANA (wt.%). Reprinted with permission from Ref. 190, Copyright © 2013 Elsevier Ltd.

Table 6.8 Average diameter

Sample Name	Avg. Diameter of Nanofibers (nm)
P0	113 ± 23
P10	105 ± 23
P15	102 ± 20
P20	94 ± 16

Source: Reproduced with permission from Ref. 190, Copyright © 2013 Elsevier Ltd.

Chapter 7

Electrochemical Capacitive Behavior of Nanostructured Conjugated Polymers

7.1 Electrochemical Capacitive Behavior of Nanostructured Conjugated Polymers

Conjugated polymers (CPs) in doped form have been used as energy storing materials in electrochemical capacitors due to their large pseudocapacitances. CPs become doped during the redox process in which the charges are stored. Due to high surface areas and small dimensions, conjugated polymeric nanostructures are of different importance in the production of capacitors, which allow diffusion of electrolytes in the polymer matrices and more available active sites for redox reactions. Additionally, the double-layer capacitance of a nanostructured CP film is also much larger than that of a compact one. CP-based capacitor can store charges in its whole volume, so it can reach a much higher specific charge density in comparison to carbon materials.[192]

Electropolymerization is an important technique to prepare CP-based capacitors which control the thickness and morphology of CP nanomaterials, for instance, CP/carbon nanocomposites are considered as good candidates for capacitors. In these nanocomposites, i.e., carbon nanotube (CNT), active carbon

Nanofibers of Conjugated Polymers
A. Sezai Sarac

ISBN 978-981-4613-51-4 (Hardcover), 978-981-4613-52-1 (eBook)
www.panstanford.com

nanoparticles behave as templates for the electrosynthesis of CPs, and they also increase the conductivity of the composite electrode and ease electron transportation in the device. Some electroactive inorganic materials can also be blended into CP matrices to prepare capacitors. Electrosynthesized polyaniline (PANI) nanocomposites were studied indicating high specific capacitances due to the combination of the capacitances of both components.[192]

7.1.1 PANI Electrochemical Supercapacitor

PANI is one of the promising materials in supercapacitors due to its high capacitive characteristics, low cost, and easy synthesis. The materials in nanosize form with high surface area and high porosity give the best performance as electrode materials for supercapacitors because of their distinctive characteristics of conducting pathways, surface interactions, and nanoscale dimensions. The synthesis and capacitive characterization of high-surface-area nanomaterials, that is, nanotubes, nanowires, etc., have been carried out extensively in the past decades. Consequently, different indirect methods were used to synthesize nanosized PANI, such as template synthesis, self-assembly, emulsions, and interfacial polymerization.[193]

PANI was potentiostatically deposited on to 1 × 1 cm stainless-steel (SS) plates (SS) (grade 304, 0.2 mm thick) at 0.75 V versus a saturated calomel electrode (SCE). An electrolyte solution of 1 M H_2SO_4 + 0.05 M PANI was used for the electrochemical deposition of PANI nanowires on the SS electrodes. Figure 7.1 shows the scanning electron microscopy (SEM) images of PANI nanowires formed after 10 min deposition time. The diameter of the nanowires is in the range of 30–60 nm. The thickness and size of the nanowire film are ~20 μm and 1 × 1 cm, respectively. The nanowire's network is highly porous with interconnectivity. A possible growth process of such nanowires is proposed, as shown in Fig. 7.1a,b. The PANI nanowires grow via the seedling growth process, in which further PANI is deposited on the initially deposited nanosized granules. As the deposition progress, an aligned nanowire network is formed. Further deposition results in extended length and in the misalignment of nanowires, and crosslinks are formed.[193]

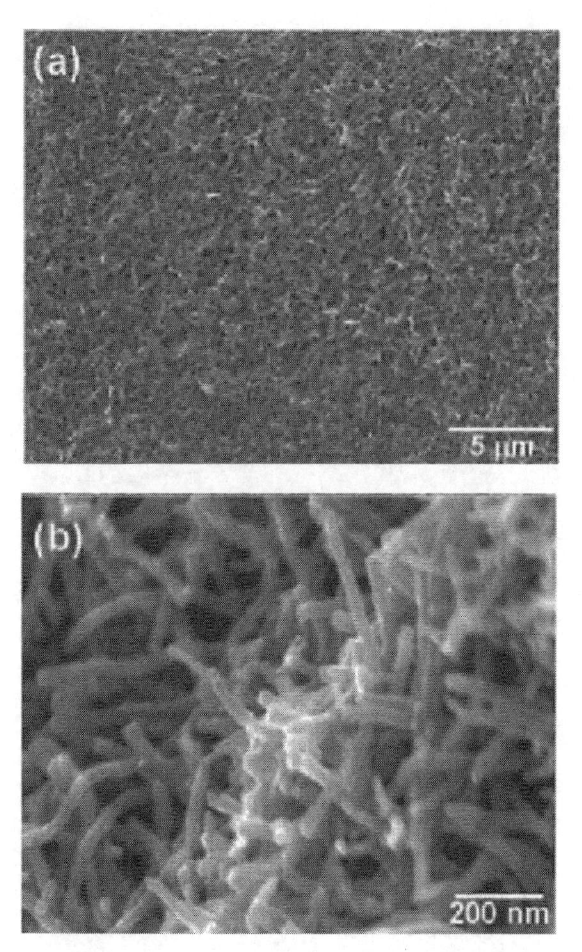

Figure 7.1 (a–b) SEM images of polyaniline nanowires at different magnifications. Reprinted with permission from Ref. 193.

Figure 7.2a shows the X-ray diffraction spectra of PANI nanowires. A very broad peak was observed centered at $2\theta \approx 19°$ with a shoulder at $2\theta \approx 26°$. This suggests that the nature of the nanowires is amorphous. Figure 7.2d shows the ultraviolet-visible (UV-Vis) spectra of the PANI nanowires in the dedoped state in *N*-methyl-2-pyrrolidone (NMP) solution. Before obtaining the spectra, the PANI nanowires were washed in 0.1 M NH_4OH solution. In this process, original green PANI becomes blue-purple, indicating

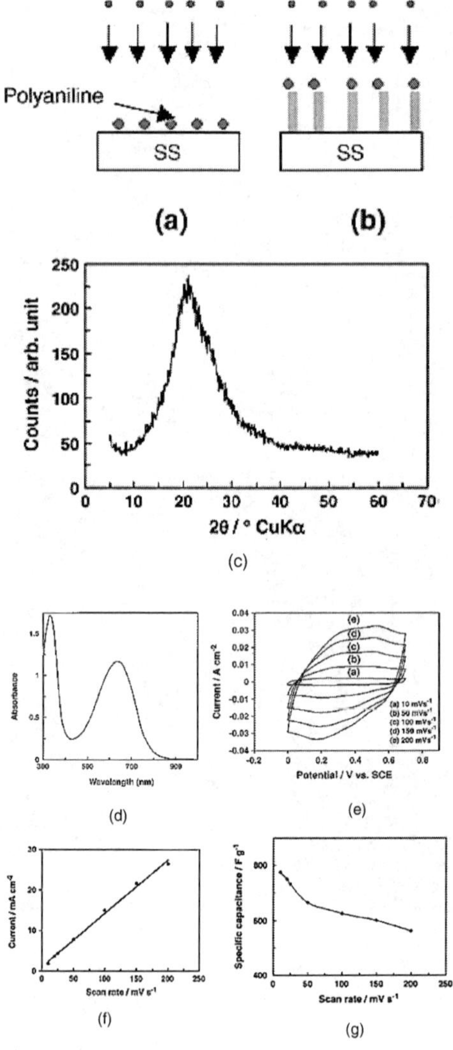

Figure 7.2 (a–b) Schematics of the possible growth process of polyaniline nanowires. (c) X-ray diffraction pattern of the original polyaniline nanowires. (d) UV-Vis-NIR spectra of polyaniline nanowires in the dedoped state in NMP solution. (e) Cyclic voltammograms of polyaniline nanowires at different sweep rates in 1 M H_2SO_4 electrolyte. (f) CV current density of polyaniline nanowires vs. the scan rate and (g) specific capacitance of polyaniline nanowires calculated from CV vs. the scan rate. Reprinted with permission from Ref. 193, Copyright © 2005 Elsevier B.V. All rights reserved.

that the doped emeraldine form has been deprotonated to emeraldine base. A few drops of this base formed a clear blue solution in the NMP, giving UV-Vis-NIR absorbance spectrum with maxima at 638 and 332 nm, in good agreement with the literature results for indirectly prepared PANI nanostructures.[193]

Figure 7.2e shows a cyclic voltammogram (CV) of PANI nanowires/SS electrode in 1 M H_2SO_4 electrolyte in the potential range of 0 and 0.7 V versus an SCE. The near rectangular-shaped CVs at a low scan rate and high overall current suggest the highly capacitive behavior of the PANI nanowires. The CVs were also recorded in the same electrolyte with a bare SS electrode, but no capacitive behavior was observed. Hence the capacitive behavior of PANI nanowire/SS electrode is entirely due to PANI nanowires.[193]

The CV current densities and the calculated specific capacitance value (C_{sp}) of 775 F/g^{-1} were obtained at a scan rate of 10 mV/s, whereas a C_{sp} value of 562 F/g^{-1} was obtained at a high scan rate of 200 mV/s. This decrease of 25% in the specific capacitance at high scan rates is much lower than in the case of metal oxides where a decrease of 50%–80% was reported between 10 and 200 mV/s. Moreover the CV current increases linearly with an increase in the scan rate (Fig. 7.2f). This implies highly stable supercapacitive characteristics of the PANI nanowires.[193]

Figure 7.2g shows the cyclic stability of PANI nanowires at a sweep rate of 100 mV/s for 1500 cycles. There is a small decrease in the specific capacitance value in the first 100 cycles and thereafter the specific capacitance remains almost constant. The decrease in the specific capacitance was 8% in the first 500 cycles and 1% in the subsequent 1000 cycles, indicating high stability of the electrode for a long cyclic life.[193]

7.2 Polypyrrole and Its Composites

The principle of electroconduction in polypyrrole (PPy), similar to other conjugated polymers, depends on the carriage of electrical charge in the polymer chain by the movement of the bipolaron (dication) and polaron (radical cation) produced within the backbone of the polymer by the oxidation of the chain (Figs. 7.3 and 7.4).

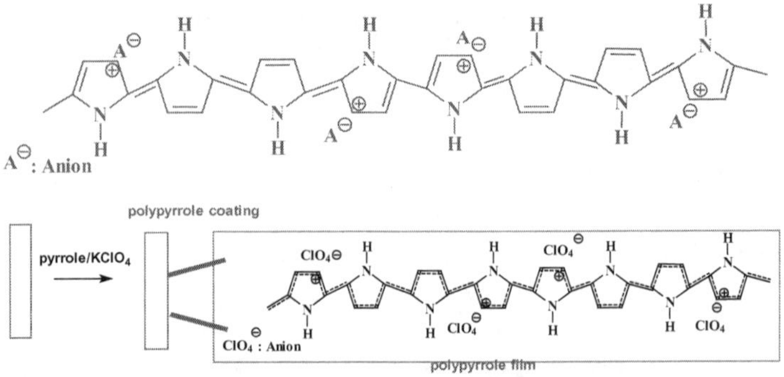

Figure 7.3 Electrochemical deposition of polypyrrole.

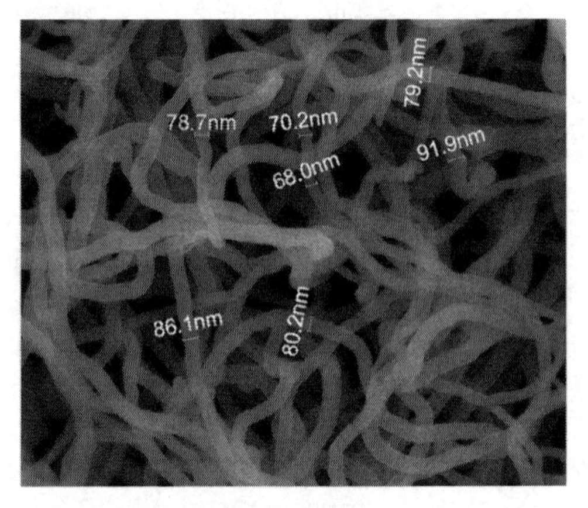

Figure 7.4 Polypyrrole nanowires have been deposited on interdigitated transducers through template-free electropolymerization. Reprinted with permission from Ref. 194, Copyright © 2012, American Chemical Society.

Polypyrrole was formed at a constant current until 4.5 C/cm^2 of current density was achieved from 0.15 M pyrrole in 0.1 M $KClO_4$ solution. The nanowires formed were 40–90 nm in diameter according to SEM analysis, and some of them were bridging the insulating gaps between gold electrodes.

An X-ray photoelectron spectroscopy study has been conducted to determine the chemical composition of the synthesized nanomaterial.

The developed sensors were tested toward different concentrations of hydrogen gas at room temperature, and their sensitivities were compared.[194]

7.2.1 Synthesis of Polypyrrole and PEDOT Carbon Nanofiber Composites

Nanosize-thin PPy films can be deposited on vapor-grown carbon fibers (VGCFs) by using an in situ chemical polymerization of the monomer in the presence of $FeCl_3$ oxidant. This is a well-known method to obtain thin conjugated polymeric films. An ultrasonic cavitational stream was used during polymerization of pyrrole (Py) to enable the deposition of uniformly nanothin PPy films on the surface of VGCF.[195]

The Fourier transform infrared (FTIR) spectrum of the PPy/VGCF composites (KBr pellet) showed that the PPy bands (conjugated double bond: 1548 cm^{-1}; amines: 1172 cm^{-1}, etc.) are present in the spectrum of the composite electrode, indicating that PPy had deposited on VGCF.

Figure 7.5 shows SEM micrographs of the surface of VGCF after chemical deposition and polymerization of Py with different molar concentrations.[196] The surface becomes more uniformly smooth with decreasing the molar concentration of Py in 20% aqueous methanol solution. At a higher concentration of the Py, especially at 0.75 M, considerable amount of agglomerates grown on the surface of VGCF were observed. The thickness of PPy layer on VGCF could be observed, as shown in Fig. 7.6 (transmission electron microscopy [TEM]) having different monomer content of (a) 0.2 M, (b) 0.1 M, (c) 0.05 M, and (d) 0.02 M pyrrole. It was observed that the PPy layer becomes thinner with decreasing monomer concentration in the initial polymerization solution.

> Liquid crystalline graphene oxide was used as precursor to interact with poly(3,4-ethylenedioxythiophene):poly(styrenesulfonate) (PEDOT-PSS) in dispersion in order to form a conductive polymer entrapped, self-assembled layer-by-layer structure. This layer-by-layer self-assembled 3D composite (reduced graphene oxide–PEDOT-PSS) showed better electrochemical performance of 434 F g^{-1} through chemical treatment. An asymmetric supercapacitor device using aqueous electrolyte was also studied of this same composite. The resulting performance from this set up included a specific capacitance of 132 F g^{-1}. An increase in specific capacitance with increase in cycle life emphasize the stability of such device.*

---

*Excerpt reprinted with permission from Ref. 197, Copyright 2014, Islam, Chidembo, Aboutalebi, Cardillo, Liu, Konstantinov and Dou.

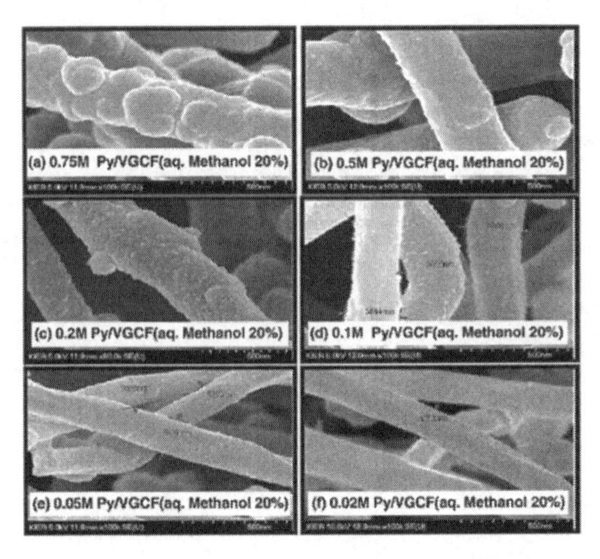

Figure 7.5 SEM images of PPy-coated VGCF (a–f) in 20% aqueous methanol using different monomer concentrations. (The mole contents are marked in the images.) Reprinted from Ref. 196, Copyright 2009, ESG.

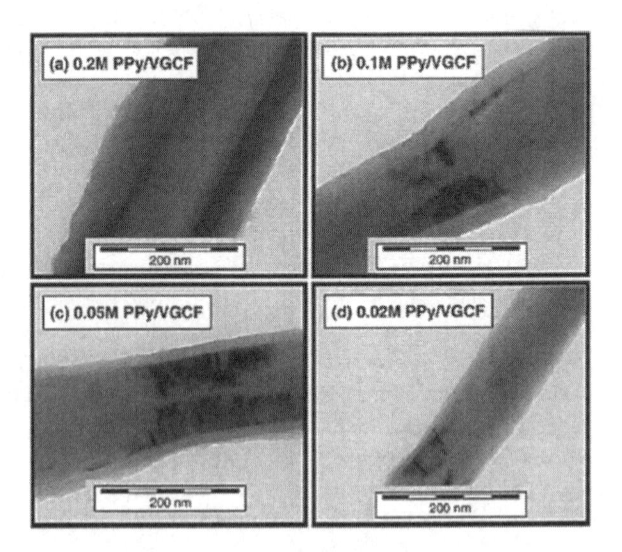

Figure 7.6 TEM images of PPy-coated VGCF having different monomer content of (a) 0.2 M, (b) 0.1 M, (c) 0.05 M, and (d) 0.02 M pyrrole. Reprinted from Ref. 196, Copyright 2009, ESG.

Cyclic voltammetry (CV) was used to determine the effects of reduced graphene oxide in the rGO–PEDOT:PSS composite as shown in Fig. 7.7. An increase in the area of the CV suggests that the reduction of GO has a massive impact on the overall specific capacitance of the composite. From the same plot, the EDLC and pseudocapacitive contributions from the rGO and PEDOT-PSS are evidenced by the large current separation and distorted rectangular shape of the CV. It is assumed that the overall pseudocapacitive effect arises from both the rGO and the PEDOT-PSS.*

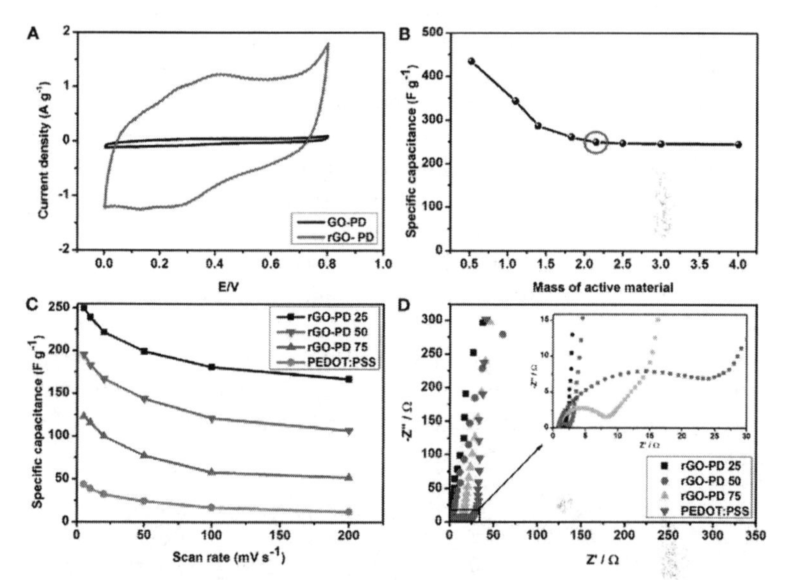

Figure 7.7 Electrochemical performance of rGO–PEDOT:PSS composite in a three-electrode system: (A) cyclic voltammograms of the composite electrodes before and after chemical reduction, (B) specific capacitance of rGO–PEDOT:PSS composite (rGO PD 25) composite electrodes with different mass loading at 5 mV s^{-1}, marked with the minimum active material mass for further electrochemical characterization, (C) variation of specific capacitance of different composites including PEDOT-PSS with scan rates, and (D) Nyquist plots of different composite electrodes (increasing PEDOT:PSS content from 25 to 75%, i.e., 25% PEDOT-PSS containing dispersion of GO is indicated as GO-PD 25 composite). The inset shows the corresponding magnified high frequency region. Reprinted with permission from Ref. 197, Copyright © 2014 Islam, Chidembo, Aboutalebi, Cardillo, Liu, Konstantinov and Dou.

*Excerpt reprinted with permission from Ref. 197, Copyright 2014, Islam, Chidembo, Aboutalebi, Cardillo, Liu, Konstantinov and Dou.

7.2.2 Synthesis and Pseudocapacitance of Chemically Prepared PPy

The effect of the solvent used in the chemical polymerization of Py on the electrochemical pseudocapacitance properties of the material is investigated by means of cyclic voltammetry. Electroconducting PPys are prepared in the presence of various solutions such as H_2O, diethylether, and acetonitrile (ACN).

To determine the electrochemical properties of PPy, cyclic voltammetry was used for ACN containing 1 M tetraethylammonium tetrafluoroborate ($TEABF_4$). The potential was scanned in the range −1 to +1 V and the scan rate was fixed at 5 mV/s.

Scanning electron micrographs of the PPy powder prepared with different solvents indicated that the degree of surface roughness of the PPy powder increases with the solvent used in the following order: H_2O > diethylether > ACN. The surface of PPy powder prepared in ACN solvent has a much denser structure than that of the powder obtained in either H_2O or diethylether. Since the morphology of PPy strongly depends on the solvent used during polymerization, it can be concluded that PPy solubility and solvent polarity have a significant role in this behavior. These findings are in agreement with previous results and show that the morphological properties of PPy powder can be controlled through choice of the solvent.[198]

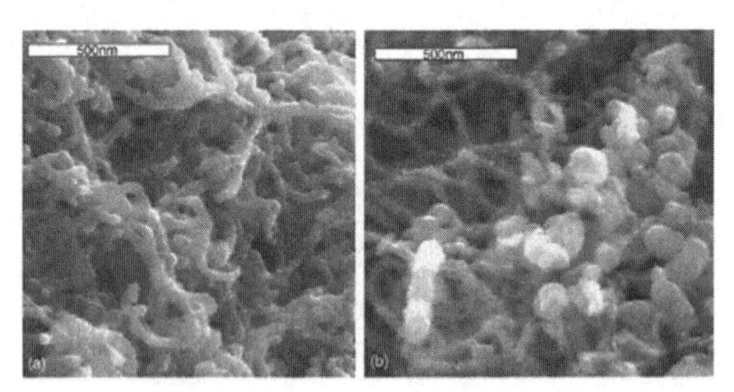

Figure 7.8 SEM micrographs showing the morphology of (a) PANI/MWCNT and (b) PPy/MWCNT composite materials, which contain 80 wt.% of conducting polymer. Reprinted with permission from Ref. 199, Copyright © 2004 Elsevier Ltd.

The specific capacitance of PPy/MWNT and PPy powder in different solvents are summarized in Table 7.1.

Table 7.1 Specific capacitance of PPy/MWNT and PPy powder

	Capacitance, (F/g)	Electrolyte	Ref.
PPy/MWNT in H_2O	320	1 M tetraethylammonium tetrafluoroborate ($TEABF_4$).	199
PPy/MWNT in ACN	55	1 M $TEABF_4$	199
PPy/MWNT in Diethylether	151	1 M $TEABF_4$	199
Polypyrrole powder	355	1 M H_2SO_4	198

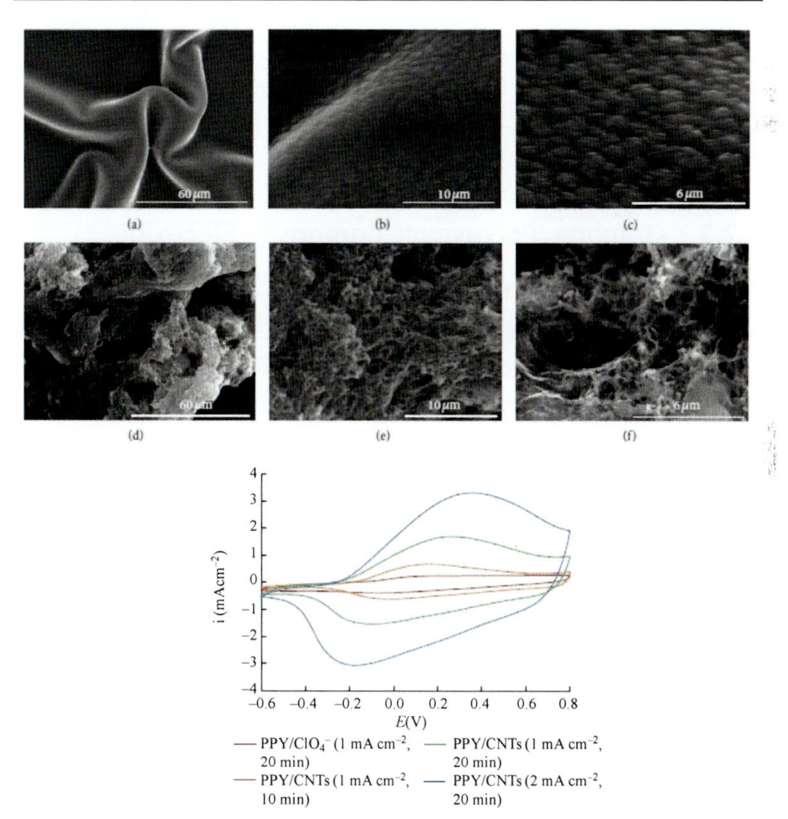

Figure 7.9 SEM morphology for PPy (a–c) and PPy/CNTs composite films (d–f) obtained galvanostatically at 0.1mA cm^{-2} for 20 min and CVs in 0.1 M $LiClO_4$ in propylene carbonate of the PPy/CNTs composite films compared with the pure polymeric ones galvanostatically obtained at different charge densities. Reprinted from Ref. 201.

The pseudocapacitance values were calculated from the voltammograms and it is found that the PPy powder prepared in the H_2O electrochemically more reversible than the powder prepared in ACN or diethylether solution, as demonstrated by the well-defined shape of the voltammogram. This result means that the ion diffusion for charge compensation during the redox reaction is much faster in the porous structure of PPy prepared in H_2O than in the dense structure of PPy prepared in either diethylether or ACN. The specific capacitance of PPy prepared in H_2O is about 355 F/g^{-1}.[198]

The specific capacitance of PPy powder decreases for samples prepared in H_2O, diethylether, and ACN, respectively. The PPy prepared in H_2O has a capacitance of about 355 F/g^{-1} in ACN solution containing 1 M $TEABF_4$. This dependency on the solvent that is employed is related directly to the surface roughness of the resulting PPy.[198]

7.2.3 Graphite/Polypyrrole Composite Electrode

In the electric double-layer capacitor, energy storage arises mainly from the separation of electronic and ionic charges at the interface between electrode materials with a high specific area (such as carbon) and the electrolyte solution. In the redox capacitor, fast faradaic reactions take place at the electrode materials at characteristic potentials, as in batteries. While activated carbon has demonstrated a higher cycle life, its capacitance value and electrical conductivity are lower than those of CPs. On the other hand, electrically CPs including PPy, PANI, etc., have been found to have higher capacitance because capacitive and faradaic currents contribute to the charge storage. The CPs have disadvantages that include a lower cycle life in charge–discharge duty than carbon-based electrodes because the redox sites in the polymer backbone are not sufficiently stable for many repeated redox processes.[200]

Some methods have been attempted to improve the capacitance and conductivity of carbon-based supercapacitors. For example, fabrication of a CP/graphite fiber composite with PPy has been done by using electrochemical polymerization of Py on the surface of graphite fiber and shows higher capacitance and conductivity.[200]

The electrochemical method has two disadvantages compared to the method of chemical polymerization. One is that the electrochemical method is limited in terms of the mass production of composite electrodes. The other disadvantage is that the electrochemical method is not suitable for preparing controlled polymer films with thicknesses above 100 μm, although it is suitable for preparing very thin films of polymer. If a thin polymer film with a very high specific capacitance per unit mass is converted to specific capacitance per unit area, this value is less than that for a carbon-based supercapacitor because of the small deposited polymer mass.[200]

PPy/graphite composite electrodes fabricated chemically by using sequential dipping methods, can be advantageous compared to the electrochemical composite method.[200]

The total capacitance per unit area of the electrode (at 10 mV/s) increased monotonically with cycle time dipping increasingly, but this behavior became saturated at four to five times, as shown in Fig. 7.10. It was found that the current response becomes increasingly influenced by the capacitive time constant (RC) of the sample at higher sweep rates. On increasing the sweep rate, the response is dominated by the RC time constant such that a capacitance region is not observed.[200]

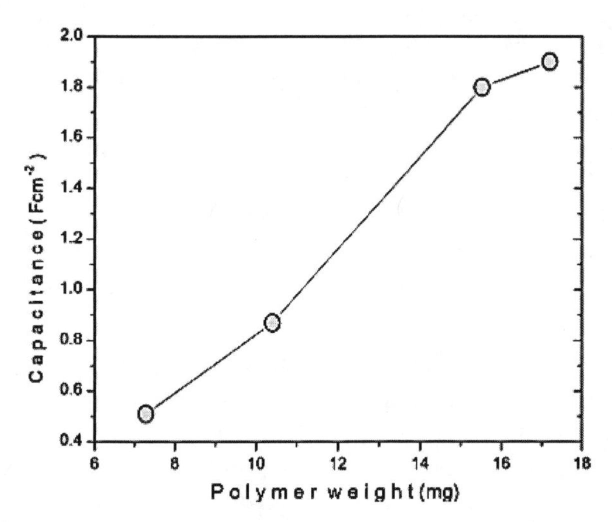

Figure 7.10 Capacitance of PPy/graphite fiber composite electrode in 1 M KCl at a 10 mV/s sweep rate. Reprinted with permission from Ref. 200. Copyright © 2002 Elsevier Science B.V.

A PPy-coated graphite composite prepared by the chemical dipping method has been suggested to be suitable for the manufacture of supercapacitor energy storage devices. The sequential dipping method provides a well-coated, thickness-controlled PPy/graphite fiber composite electrode. The electrodes show a specific capacitance of ~400 F/g and a coulombic efficiency of 96%–99%. This demonstrates that the dipping method is suitable for the preparation of a composite electrode for supercapacitors.[200]

7.3 EIS Study of Poly(3,4-Ethylenedioxythiophene)

Poly(3,4-ethylenedioxythiophene) (PEDOT) belongs to a group of very stable CPs that are potential candidates for many technical applications, including antistatic coatings and solid electrolyte capacitors, electrochromic devices, biosensors, and all-solid-state ion sensors. PEDOT is electroactive in aqueous solutions exhibiting a stability superior to that of PPy. Furthermore, ion diffusion in PEDOT contacted by a polymer electrolyte was about 3 orders of magnitude faster than for other conjugated polymers. The electrochemistry of PEDOT in more detail by electrochemical impedance spectroscopy (EIS) is challenging; EIS is a powerful technique to study charge transfer, ion diffusion, and capacitance of CP-modified electrodes.[201,202]

EIS was used to study the charge transfer, ion diffusion, and capacitance of PEDOT films doped with small mobile anions (Cl^-) or large immobile polyanions (polystyrene sulfonic acid [PSS^-]), resulting in PEDOT films with anion and cation exchange behavior, respectively. The PEDOT films were studied in contact with aqueous solutions containing different anions (Cl^-, PSS^-) and cations (K^+, Na^+). The good stability of PEDOT allows accurate characterization of its electrochemical properties without any significant degradation of the material.[203]

Typical impedance spectra of the Pt:PEDOT(KCl) electrode in 0.1, 0.025, and 0.00625 M KCl are dominated by a 90° capacitive line, which extends down to very low frequencies (0.01 Hz) for thick films of PEDOT.[203]

At high frequencies, there is only a slight deviation from the capacitive line, indicating fast charge transfer at the metal–polymer and polymer–solution interfaces, as well as fast charge transport in the polymer bulk. The high-frequency intersection with the Z_{re} (or Z') axis depends strongly on the electrolyte concentration and is consequently determined mainly by the solution resistance and not by the ohmic resistance of the polymer film. Also, the Pt:PEDOT(NaCl) and Pt:PEDOT(NaPSS) electrodes show the same general shape of the impedance plot as the Pt:PEDOT(KCl) electrode.[203]

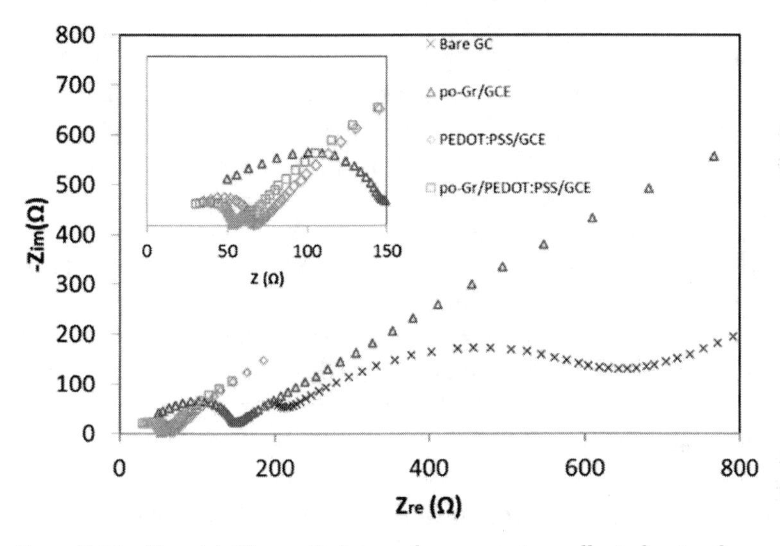

Figure 7.11 Nyquist (Z_{re} vs. Z_{im}) impedance spectra collected using bare GCE, po-Gr/GCE, PEDOT-PSS/GCE, and po-Gr/PEDOT-PSS/GCE electrodes in 2.5 mM $[Fe(CN)_6]^{4-/3-}$ and 0.1 M KCl. Inset: enlarged high-frequency region. Reprinted from Ref. 204, Copyright 2014, Yasri, Sundramoorthy, Chang, and Gunasekaran.

At the lowest frequencies studied (down to 0.01 Hz), the impedance plots deviate from the 90° line, forming the beginning of a semicircle, for a thin PEDOT film (polymerization charge = 1 mC). The appearance of the low-frequency semicircle in the impedance plots observed in the presence of oxygen is in agreement with the theory, which takes into account both ion and electron transfer between the polymer and the solution.

Partially oxidized graphene flakes (po-Gr) were obtained from graphite electrode by an electrochemical exfoliation method. As-produced po-Gr flakes were dispersed in water with the assistance of poly(3,4-ethylenedioxythiophene)/poly(styrenesulfonate) (PEDOT-PSS).[204]

Figure 7.11 is the Nyquist plot of the modified glassy carbon electrodes (GCEs) in 2.5 mM $[Fe(CN)_6]^{4-/3-}$ in 0.1 M KCl. The semicircular part in the high-frequency region represents electron-transfer-limiting process with its effective diameter equal to Faradaic charge transfer resistance (R_{ct}), which is responsible for the electron transfer kinetics of redox reactions at the electrode-electrolyte interface. The modified electrodes, po-Gr/GCE, PEDOT-PSS/GCE, and po-Gr/PEDOT-PSS/GCE, had R_{ct} values 144 Ω, 65 Ω, and 54 Ω, rspectively, that were lower than that for bare electrode (R_{ct} = 228 Ω), which indicated that higher conductivity can be obtained as a result of modification processes.Generally poor conductivity of GO-modified electrode is due to the presence of excessive oxygenated species, which accentuates the insulating characteristics. The R_{ct} for po-Gr film is also higher than for PEDOT-PSS/GCE and po-Gr/PEDOT-PSS/GCEs due to the presence of oxygenated species associated with po-Gr, which may affect conductivity of the electrode. However, when po-Gr sheets present in PEDOT-PSS, it improves the conductivity of the electrode (Fig. 7.11).[204]

7.3.1 Carbon Nanofibers and PEDOT-PSS Bilayer Systems

High-surface carbon materials have been considered as electrode materials for electrochemical capacitor cells. The great diversity of carbon (as powders, fibers, paper, nanotubes, nanofibers, aerogels, and nanocomposites) gives many possibilities to the whole field of research. Furthermore, carbon has a relatively low cost, no toxicity, high chemical stability in various solutions, and high thermal stability. Introduction of electroactive metallic particles or CPs has been carried out to modify the pore size distribution, and asymmetric cells have been fabricated to enhance their performance. The capacitance of these materials is given basically by the double-layer storage mechanism. A larger specific surface area gives greater specific capacitance C and a greater specific energy W ($W = CV^2/2$, where V is the potential). On the other hand, the resistance due to the carbon material's porosity is associated with the power density P ($P = W/\Delta t$, where Δt is the discharge time, and $P = V^2/\Delta R$, where R is the resistance). Materials characterized by high conductivity and large pores demonstrate a high specific power.[205]

Carbon nanofibers are fibers of nano or submicron size (50–500 nm of diameter and 50–100 μm length) that have a graphitic structure. Unlike CNTs, carbon nanofibers do not have a hollow center and they have exterior walls with many edges. The presence of a wide pore size distribution is possible due to these edges, resulting in a greater surface area and high capacitance. On the other hand, conducting organic polymers such as polyacetylene, PPy, PANI, and polythiophenes can improve the energy density of electrochemical capacitors (ECs) through redox processes that contribute a pseudocapacitive storage mechanism in addition to the capacitance of the carbon material.[205]

The carbon mesoporous structure exhibits mechanical stability to the composite in a way that adapts to the polymer volumetric changes, resulting in a more stable capacitance in successive charge–discharge cycles. The electrode fabrication based on carbon nanofibers with PEDOT-PSS has been studied, bilayer and composite electrodes were electrochemically characterized in a three-electrode cell configuration, and in a symmetric and asymmetric electrochemical capacitor cell with a two-electrode configuration. Polythiophene derivative has been selected because of its stability in air and humidity, specifically PEDOT-PSS due to high electric conductivity on its *p*-doping stage and fast redox kinetics. The electrode fabrication method as well as the effect of a small amount of CPs on the capacitance values was electrochemically evaluated.[205]

Chapter 8

Preparation of Conductive Nanofibers

8.1 Preparation of Conductive Nanofibers

There are various methods to synthesize polymer nanostructures, i.e., template synthesis, chiral reactions, self-assembly, interfacial polymerization and electrospinning. Recent developments in conducting polymer nanotubes and nanofibers were summarized by Long et al.[206] Different preparation methods, physical properties, and potential applications of one-dimensional nanostructures of conjugated polyaniline (PANI), polypyrrole (PPy) and poly (3, 4-ethylenedioxythiophene) (PEDOT) were discussed.

Different routes were reported to obtain conductive nanofibers based on electrospinning. Generally, conjugated polymers may not be simply electrospun due to their low molecular weights, poor solubility and rigid backbone structure. Different techniques, i.e., introducing side chains, controlling main-chain architecture, synthesizing new monomers, and using functional dopants have been applied. To solve the processability problems of conducting polymers, introduction of alkyl groups into monomer have been used.

Blending with other polymers and coating insulating polymer with conducting polymers are the most familiar techniques.

Nanofibers of Conjugated Polymers
A. Sezai Sarac

ISBN 978-981-4613-51-4 (Hardcover), 978-981-4613-52-1 (eBook)
www.panstanford.com

Blending with a spinnable polymer is a common way to compensate poor spinnability.[207] However, the presence of an insulating carrier polymer introduces a conductivity percolation threshold by limiting their usage in the applications where high conductivities are required. The polymerization of a conductive monomer on the surface of a fiber, made with a common polymer and a catalyser/doping agent is also another approach.[208,209]

8.2 Polyaniline Nanofibers

The emeraldine salt of polyaniline which is one of the three different forms of polyaniline—leucoemeraldine (the fully reduced state), emeraldine (the half oxidized state) and pernigraniline (the fully oxidized state) (Fig. 8.1)—is electrically conductive while the others are insulators. Conventional chemical synthesis of polyaniline is based on an oxidative polymerization of aniline by using an oxidant in the presence of a strong acid dopant in the reaction medium.[1]

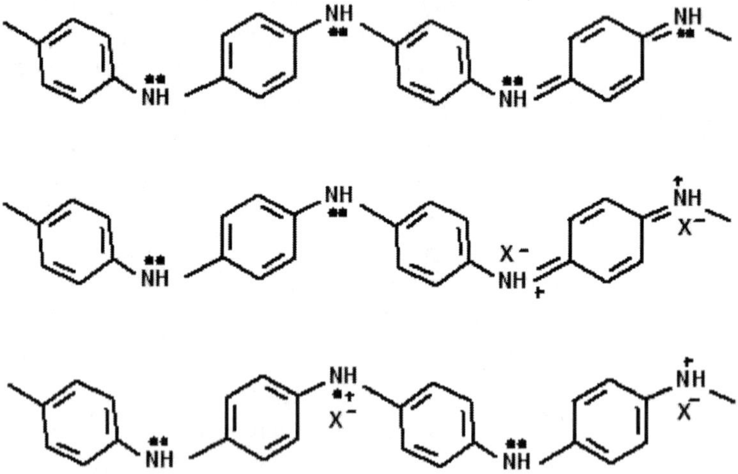

Figure 8.1 Polyaniline is doped with acids (X^- represents the counterion group).

Due to low cost, simple way of synthesis, high stability at room temperature, good optical and electrical properties polyaniline has drawn considerable attention for electronic and optical devices, sensors, light-emitting diodes, rechargeable batteries and gas separation membranes. On the other hand, processing polyaniline is limited as it is highly rigid polymer due to its chemical structure made of reduced and oxidized repeating units and aromatic rings, with intermolecular hydrogen bonds and charge delocalization.[210–212]

Nanofibers of polyamide 6 (PA6) with different contents of poly (aniline) (PANI) doped with *p*-toluene sulfonic acid (TSA) were obtained by blending method. [1]

MacDiarmid's group succeeded to produce pure PANI fibers by electrospinning. According to the authors, 100% PANI fibers with an average diameter of 139 nm and a conductivity value of a single fiber, ~0.1 S/cm, were produced placing a 20 < wt.% solution of PANI in 98% sulfuric acid in a glass pipette above a copper cathode immersed in pure water at a 5000 V potential difference.

To improve PANI processability, the first approach to obtain polyaniline doped with camphorsulfonic acid (PANI-CSA) and doped nanofibers blended with common polymers by electrospinning was done by MacDiarmid's group. In another study a nonwoven mat was obtained by using a PANI/poly(ethylene oxide) (PEO) solution dissolved in chloroform. By controlling the ratio of PANI to PEO in the blend, fibers with conductivity values comparable to that of PANI-CSA/PEO cast films were produced.

The production of PANI(CSA) nanofibers dispersed in a poly(methyl methacrylate) (PMMA) solution in chloroform was realized and aligned fibers with diameters in the range of 500 nm to 5 μm were obtained and direct current (DC) conductivity was estimated to be around 0.28 S/m.

A mixture of dimethyl sulfoxide/tetrahydrofuran (THF) was used to obtain homogeneous blended nanofibers of an HCl-doped poly(aniline-co-3-aminobenzoic acid) (3ABA-PANI) copolymer and poly(lactic acid) (PLA) for tissue engineering. Since the solvent system dimethyl sulfoxide (DMSO)/THF (50:50) is quite difficult to be removed from the nonwoven mats, the authors used a heated collector to facilitate solvent removal. This procedure can be considered essential in cases where nonwoven fiber mats will be used in cell growth. Besides, composite electroactive fibers achieve lower conductivity values compared to pure conjugated polymer fibers; this lower conductivity is quite appropriate for tissue engineering.

As observed earlier, the addition of ions to the electrospinning solution can improve fiber spinnability due to the increase of charge carriers in the solution. Conjugated polymers may have a high density of charge carriers and this can also affect the electrospinning process. The addition of polyaniline and p-toluene sulfonic acid (PANI.TSA) to a PLA solution in hexafluoropropylene (HFP) caused a similar effect of inorganic salts' addition, and the average fiber diameter reduced to ~400 nm after addition of 0.2 <wt.% of PANI.TSA. Also, the diameter distribution narrowed and a beadless morphology was observed.

Furthermore, electro-rheological effects can be observed in the polymer solution with a high density of charge carriers.

Composite nanofibers of poly(vinylidene fluoride trifluoroethylene)/PANI–polystyrene sulfonic acid (PSS) with diameters of ~6 nm were reported.

The addition of the conjugated polymer PANI and polystyrene sulfonic acid (PSSA) also increased the charge density of the solution and assisted the fabrication of homogeneous nanofibers at lower than normal poly(vinylidene fluoride) (PVDF) concentrations in DMF.

To produce aligned PANI-based nanowires and nanotubes based on electrostatic steering, electrospun nanofibers can be aligned on a substrate using an alternative electrostatic field generated between two collectors. This technique suggests strategies to achieve fiber alignment and counting with an "immobile" experimental setup.

Conducting nanofibers by blending multiwalled carbon nanotubes (MWCNTs) and PANI/PEO were produced using electrospinning.

The unexpected transition in the electrical conductivity was attributed to the interactions between the MWCNTs and the CPs inside the fiber due to an annealing effect of the PANI/PEO matrix from the thermal dissipation of the carbon nanotubes (CNTs). It was also related to the self-heating effect of the MWCNTs incorporated into the CPs, which will be very helpful in enhancing the electrical properties of nanoscale conducting composite fibers.

A conductive composite fibers based on MWCNTs and nylon-6,6 by electrospinning was also prepared by stable dispersions of MWCNTs functionalized with $-NH_2$ terminations in formic acid.

Afterward, nylon-6,6 solution in formic acid was electrospun with different filler concentrations.

Increase in conductivity can be attributed to the enhancement of the electron conduction process by the increase of MWCNT content.

Electrospun nanofibers of two series of binary blends of poly[2-methoxy-5-(2-ethylhexyloxy)-1,4-phenylene vinylene] (MEH-PPV) with regioregular poly(3-hexylthiophene) (P3HT) (with diameters of 100–500 nm) and (MEH-PPV) with poly(9,9-dioctylfluorene) (PFO) was studied. Phase-separated morphology was observed by scanning electron microscopy (SEM) images. Due to confinement of the liquid jets during electrospinning, the length scales of the phase separation in these blend fibers are much smaller than those of the MEH-PPV/P3HT blend thin films prepared by spin coating where the length scales of the phase-separated domains were on the order of 100–150 nm.[213]

Furthermore, the red shift in electronic absorption peaks suggests that the polymer chains in the fibers are more extended, which may lead to the increase of the π conjugation length.[213]

Moreover, the extended polymer chains should be oriented along the fiber axis due to the strong stretching of the liquid jet during electrospinning better π electron delocalization. In the absorption spectra of MEH-PPV/PFO blend nanofibers a 20–30 nm red shift of the PFO absorption band to 400–410 nm was observed, suggesting that the PFO chains are also extended and oriented along the fiber axis.[213]

In the PANI.TSA/PLA blended electrospun nanofibers no phase segregation of PANI in a PLA matrix was observed, while phase segregation was observed in cast films with the same composition. Due to rapid solvent evaporation in the electrospinning process, no crystalline structures in fiber mats were formed compared to cast films. Highly homogeneous electroactive fibers can be useful in the construction of electronic devices and sensors. Similar behavior was observed in the PVDF-TrFE/PANI-PSSA electrospun nanofibers.

The acetone bath was used for the formation of pure polyaniline fibers by electrospinning method.[214] The fibres were collected in an acetone bath placed on the collector electrode. Because by placing the acetone bath make the excess of solvent to diffuse into the acetone allowing the pure polyaniline fiber formation. In another study, PANI/silica hybrid nanofiber webs were prepared by using two methods.[215] The first method was in situ polymerization of aniline-doped silica nanofibers in ammonium persulfate solution. The second method was immersing the silica nanofibers into the solution of PANI.

Polyethylene oxide (PEO) was blended with camphor-10-sulfonic acid (CSA) doped conductive PANI by Neubert et al.[216] Camphorsulfonic acid doped polyaniline (PANCSA) blends with polyethylene oxide (PEO) were studied. The conductive CSA/PANI-PEO composite fibers have been used as the conductive collector for electrospraying process. Titanium dioxide (TiO_2) nanoparticles were sprayed and adsorbed on the fibers. The degree of adsorption and dispersion of nano TiO_2 catalysts on the surface of the fibers depended on weight percentage (wt %) of PANI in PEO solution and the strength of electrical conductivity of the fibers used during electrospraying.[1]

PANI was blended with a natural protein, gelatin, and co-electrospun into nanofibers for the application of such a blend as a conductive scaffold for tissue engineering purposes.[217] The doping of gelatin with a low content PANI leads to an alteration of the physicochemical properties of gelatin. To test the usefulness of PANI–gelatin blends as a fibrous matrix for supporting cell growth, H9c2 rat cardiac myoblast cells were cultured on fiber-coated glass cover slips. Cell cultures were evaluated in terms of cell proliferation and morphology (Fig 8.2). Figure 8.2 shows SEM micrographs of gelatin fibers electrospun from pure gelatin solution and the various PANI-gelatin blends. With increasing the concentration of PANI in the solution, the fiber sizes decreased. Electrospun nonwoven webs were obtained from solution of poly(3-hydroxybutyric acid) (PHB) and dodecylbenzene sulfonic acid (DBSA) doped polyaniline in chloroform/trifluoroethanol mixture.[218] PANI nanofibers were prepared by using different concentrations (from 10.6% to 19.1%) of PANI in hot sulfuric acid solution.[219]

Nylon-6 electrospun fiber webs by the in situ polymerization of polyaniline were prepared.[220] PANI-nylon 6 electrospun fiber web with various PANI and nylon contents in a formic acid solution were prepared, when the concentration of PANI nanoparticles was from 2 to 8 wt %, the PANI-nylon 6 electrospun nanofibers were composed of two kinds of phases. When the concentration of PANI nanoparticles was over 12 wt %, the PANI-nylon 6 electrospun nanofibers had only one-type phase but defects were observed. Nonuniform morphologies might depend on the solubilities of the components in the composite structure.[1]

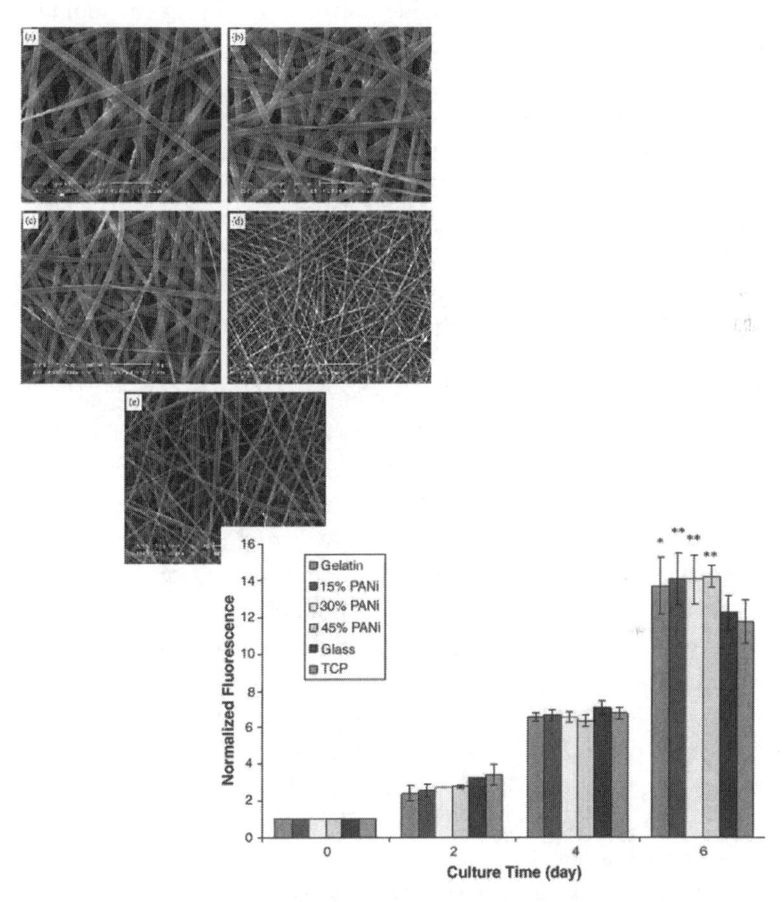

Figure 8.2 SEM micrographs of gelatin fibers (a) and PANI–gelatin blend fibers with ratios of (b) 15:85; (c) 30:70; (d) 45:55; and (e) 60:40. Original magnifications are 5000× for (a–d) and 20000× for (e). H9c2 cells were cultured on electrospun gelatin and PANI–gelatin blend fibers with volume ratio of 15:85, 30:70, 45:55, and control glass and TCP substrates over a 6 days time course. Cell proliferation/metabolic activity was evaluated using the Alamar Blue (AB) Assay (lower). Reprinted with permission from Ref. 217, Copyright © 2005 Elsevier Ltd.

8.3 Addition of Nanoparticles

The addition of nanoparticles is another way to increase the surface area and roughness of a material and alter its properties. Fabrication of biomimetic super hydrophobic fibrous mats by electrospinning of polystyrene (PS) solution in the presence of silica nanoparticles were realized. The resultant electrospun fiber surfaces exhibited a structure with the combination of nano-protrusions and grooves due to the rapid phase separation in electrospinning. The content of silica nanoparticles incorporated into the fibers proved to be the key factor affecting the fiber surface morphology and hydrophobicity. The PS fibrous mats containing 14.3 wt% silica nanoparticles showed a stable superhydrophobicity with a water contact angle as 157.2°, by approaching that (160°) of the lotus leaf. The superhydrophobicity was explained by the hierarchical surfaces increasing the surface roughness which trapped more air under the water droplets that fell on the fibers (Fig. 8.3).[221] Fang et al. applied a surface coating of silica nanoparticles and a fluorinated alkyl silane to electrospun PAN fibers. The loading of the nanoparticles and resulting surface roughness were found to be critical in generating a superhydrophobic surface. TiO_2 and graphene nanoparticles were incorporated into PS and poly(vinyl chloride) (PVC) nanofibers in order to generate self-cleaning photoelectrodes for dye-sensitized solar cells.[222] Hierarchical electrospun SiO_2 nanofibers incorporated with SiO_2 nanoparticles with fiber diameters being 500 nm and particle sizes being tens of nanometers were developed through the combination of sol–gel process and electrospinning technique followed by high-temperature pyrolysis; and their morphologies and BET surface areas were examined.

The specific surface area increased substantially upon addition of the nanoparticles from 4.14 m^2/g to 345 m^2/g with 4.29% loading. These materials are expected to find important applications in dental composites.[223]

Two-step processes to generate nanoprotrusions on electrospun fibers have also been studied. Nanofiber "shish kebabs" by electrospinning poly(ε-caprolactone) (PCL) followed by solution crystallization of PCL or a copolymer of PCL and PEO was obtained.[224]

The researchers observed a series of nanoprotrusions from spikes to blocks oriented perpendicular to the fiber axis. The mechanism for this process was similar to the crystallization mechanism of CNT growth from a surface. Carbon nanostructures from electrospun carbon nanofibers with iron and palladium nanoparticles were grown, respectively.[225,226] In a later work, the type of carbon dictated the morphology of the resulting nanostructure. Toluene as the carbon source yielded straight nanotubes, pyridine gave coiled and Y-shaped nanotubes, and chlorobenzene formed nanoribbons. Figure 8.4 displays a variety of carbon nanostructures from rods to Y-shaped protrusions. [227]

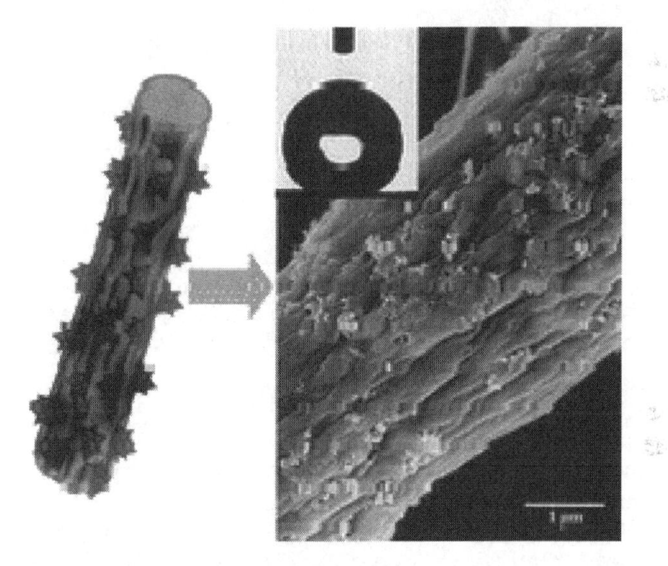

Figure 8.3 The PS fibrous mats containing 14.3 wt% silica nanoparticles showed a stable superhydrophobicity with a water contact angle as 157.2°. Reprinted with permission from Ref. 221, Copyright 2011, Royal Society of Chemistry.

8.4 Polypyrrole Nanofibers with Carriers

Due to its easy synthesis and long-term ambient stability, polypyrrole has been investigated for different applications, i.e., antistatic, electromagnetic shielding, actuators and polymer batteries. Its monomers are soluble in water. The inherently poor solubility of polymer in common solvents, which originates from the strong inter- and intra-chain interactions is the disadvantage that restricts practical applications of polypyrrole in some areas.

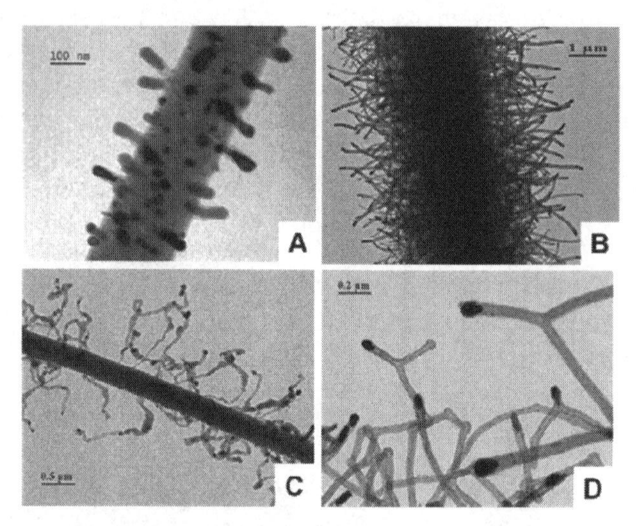

Figure 8.4 TEM images of carbon nanostructures grown on carbon nanofibers. (A) Short carbon nanotubes, (B) densely packed carbon nanotubes, (C) carbon nanoribbons, and (D) Y-shaped carbon nanotubes. From Ref. 225, © IOP Publishing. Reproduced by permission of IOP Publishing. All rights reserved.

Pure PPy was used to prepare conductive nonwoven mats. PPy was synthesized using ammonium persulfate (APS) as the oxidant and dodecylbenzene sulfonic acid (DBSA) as the dopant. Chloroform and excessive amount of DBSA were used to obtain solubility.[1] The intermolecular interaction between PPy chains were reduced by doping with high amount of DBSA but the reduction of intermolecular interaction between PPy chains decreased the interchain conduction of charge carriers and led to the decrease in bulk conductivity.

Pure (without carrier) polypyrrole conductive nanofibers were prepared by electrospinning of organic solvent soluble polypyrrole using the doping agent di(2-ethylhexyl) sulfosuccinate sodium salt (NaDEHS) and the functional doping agent NaDEHS with PEO (as carrier, Fig. 8.7) were used to obtain electrospun blends of water soluble polypyrrole.

Conductivities of nanofibers were increased up to ~10^{-4} S/cm by the increase of polypyrrole content. Morphologies and diameters of nanofibers affect properties of nanofiber mats. Polypyrrole nanofibers with the diameters in the range of about 70–300 nm were reported.[228]

The average diameter of electrospun fibers of individual fibers with and without polypyrrole are shown in SEM images (Fig. 8.5).[229] Mole percent of the initially added Py concentration varies from 0.035–0.070%. The average diameters of the nanofibers are reduced from 200 to 120 nm. In this study, elimination of surface roughness on nanofibers structure was due to the well interaction of PPy with matrix and well dispersion of latex particles in the DBSA medium resulting in an improvement in the solubility of PPy.

The relatively low molecular weight of conductive polymers, decrease in specific viscosity of composites resulted a small nanofiber diameters. Interaction of PPy with matrix creates a decrease in viscosity, and that causes the smaller diameter of nanofibers. Moreover, electrospinning solutions of nanofibers with small average diameters have exhibited higher conductivity.

Thinner fibers were obtained (70 nm) for pure (without carrier) polypyrrole nanofibers formed using $[(PPy_3)^+ (DEHS)^-]_x$ dissolved in DMF. This low average nanofiber diameter was explained by the relatively low molecular weight of the conducting polymer. The nanofibers with the average diameter of ~100 and 150 nm were obtained by electrospinning of a solution of [PPy(SO_3H)–DEHS] with 1.5 or 2.5 wt% PEO, respectively.[1]

Polypyrrole particles can be incorporated into a carrier polymer and this solution may be electrospun if materials are prepared with particulates smaller than the cross-section of the fiber. The coating process can be applied to the outer surface of a pre-spun fiber. The composite fibers of polystyrene/polypyrrole were suspended in DMF to dissolve the polystyrene leaving behind a hollow polypyrrole fiber. [1]

Long PPy fibers were obtained by a vapor deposition reaction of pyrrole on the FeAOT (an organic salt synthesized by the reaction of sodium 1,4-bis(2 ethylhexyl) sulfosuccinate (AOT) and ferric chloride) fibers. The synthesis of PPy composite fibers with multi-walled carbon nanotubes (PPy–MWCNT fibers) were reported. First, FeAOT was synthesized and then electrospun to fabricate FeAOT and FeAOT–MWCNT nanofibers. In order to produce PPy or PPy–MWCNT fibers, FeAOT or FeAOT–MWCNT fibers were placed in a reaction medium; pyrrole was deposited onto salt fibers. The PPy composite nanofibers were obtained after removing the remaining oxidant and oligomers by washing with methanol.[1]

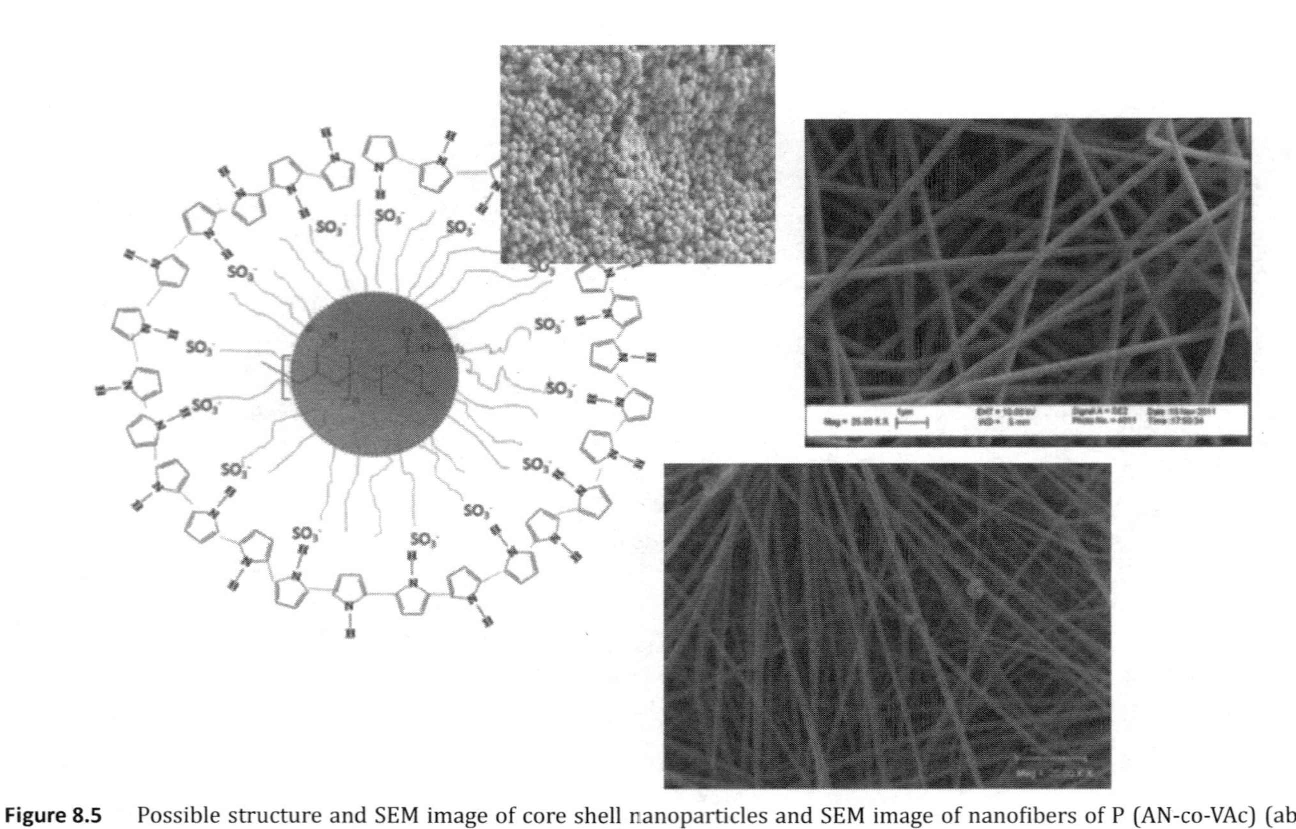

Figure 8.5 Possible structure and SEM image of core shell nanoparticles and SEM image of nanofibers of P (AN-co-VAc) (above: scale bar is 1 micron), P (AN-co-VAc)/PPy, 0.045 mole% Py (below: scale bar is 2 micron). Reprinted from Ref. 229, Copyright © 2014 by authors and Scientific Research Publishing Inc.

Silver/polypyrrole/polyacrylonitrile composite nanofibrous mats were obtained by applying coating method. $AgNO_3$/PAN mats were prepared by electrospinning and the mats were placed into the boiling mixture of pyrrole and toluene.[230] Then the pyrrole was oxidized by silver ions, leading to PPy and elemental Ag.

Figure 8.6 shows the TEM images of silver/polypyrrole/polyacrylonitrile composite nanofibrous mats prepared. It was reported that low $AgNO_3$ content could decrease the diameters of the Ag/PAN fibers and higher $AgNO_3$ concentration led to increase in the diameters of the hybrid fibers due to the high content of the solute in the electrospinning solution.[1]

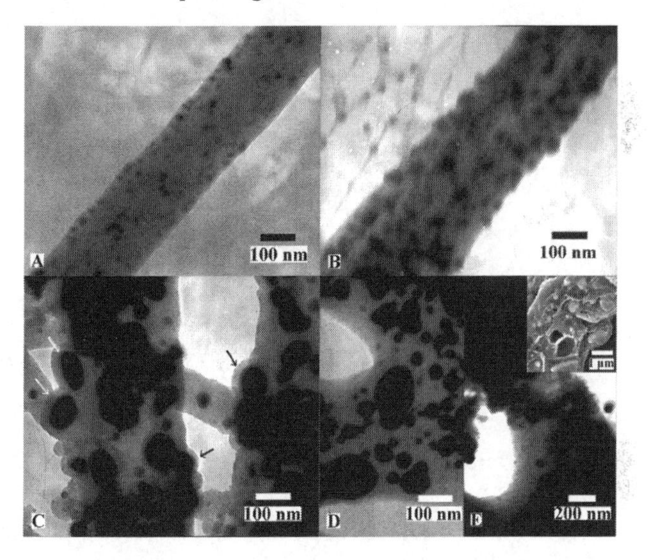

Figure 8.6 TEM images of Ag/PPy/PAN fibrous mats prepared from $AgNO_3$/PAN mats in which the content of $AgNO_3$ is (A) 10%, (B) 36%, (C) 52%, (D) 62%, and (E) 69% (inserted image: SEM image of the cross section). Reprinted with permission from Ref. 230, Copyright © 2008 Elsevier B.V. All rights reserved.

The uniform morphology is necessary for acceptable conductivities and morphology depends on polymer types, ratios of the components, and the types of solvent.[1]

Polyamide 6 (PA-6)/polypyrrole-conductive nanofibers, prepared by the polymerization of pyrrole on the fiber surface, were studied. A solution of PA-6 added with ferric chloride in formic acid was electrospun and the fibers were exposed to pyrrole vapours. Polypyrrole was formed on the fiber surface. Polypyrrole–polyethyleneoxide

(PPy-PEO) composite nanofibers were produced by Nair et al.[208] First, $FeCl_3$-containing PEO nanofibers were produced and the PEO. $FeCl_3$ electrospun fibers were exposed to pyrrole vapor for the polymerization of pyrrole. The vapor phase polymerization occurred through the diffusion of pyrrole monomer into the nanofibers. The collected non-woven fiber mat was composed of about 96 nm diameter PPy-PEO nanofibers. PEO and $FeCl_3$ were chosen due to the complexation of PEO with $FeCl_3$. $FeCl_3$ is one of the most efficient oxidants for pyrrole polymerization and chloride ions can dope PPy making it conductive. The Fe^{+3} ions are bound by the coordinating oxygen atoms of the PEO chain. This suppresses crystallization of $FeCl_3$ and assures homogeneous distribution of $FeCl_3$ along the PEO nanofibers.[1]

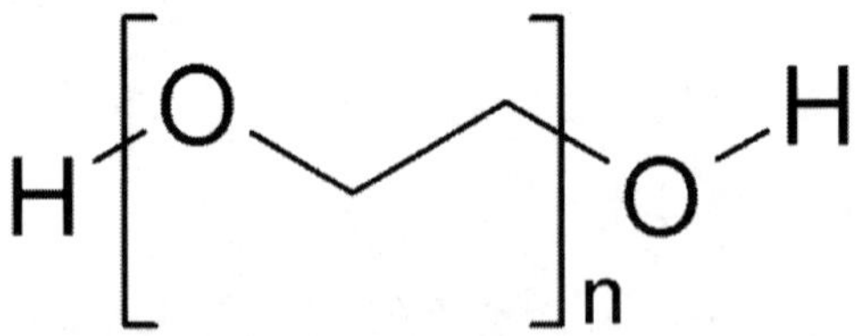

Figure 8.7 Chemical structure of poly(ethylene oxide). PEG is also known as polyethylene oxide (PEO) or polyoxyethylene (POE), depending on its molecular weight.

Core sheath conductive nanofibers were produced by coating PPy on electrospun PCL and PLA nanofibers by in situ polymerization with Fe^{3+} or APS as an oxidant, together with Cl^- or PTSA as a dopant, respectively. A nanofiber mat was immersed in an aqueous solution of pyrrole and an aqueous solution of $FeCl_3$ was added. In order to reveal the core–sheath structure after in situ polymerization, the nanofibers were soaked in dichloromethane (DCM) for 24 h to dissolve the cores. PPy layers were grown over PS nanofibers by vapor phase polymerization process. These nanofibers were produced through electrospinning of PS solutions that contained chemical oxidants capable of polymerizing pyrrole monomers.[1]

Composite nanofibers are also obtained by blending polymers. Preparation of polyacrylonitrile/polypyrrole (PAN/PPy) composite nanofibers were reported. H_2O_2 was used to polymerize pyrrole in PAN/DMF solution. Electrospun PPy/sulfonated poly[(styrene)-

(ethylene-co-butylene)-styrene] (S-SEBS) composite nanofibers were prepared. The oxidative polymerization of Py by ceric ammonium nitrate (cerium [IV]) on a poly(acrylonitrile-co-vinyl acetate) matrix and composite nanofibers by the electrospinning of their solution in DMF were demonstrated.[231,232] It was found that the conductivity (AC) of nanofibers was increased by increasing amount of PPy. PPy/PCL/gelatin composite nanofibrous scaffolds for regeneration of cardiac tissue were reported.[233]

By increasing the concentration of PPy (0%–30%) in the composite, the average fiber diameters reduced from 239 ± 37 nm to 191 ± 45 nm, and the tensile modulus increased from 7.9 ± 1.6 MPa to 50.3 ± 3.3 MPa. Nanofibers of PA6 with different amounts of PANI doped with *p*-TSA were obtained by Silva et al.[211] It was reported that as the amount of PANI increased, nanofibers diameters were increased, and higher decomposition temperatures, lower crystallinities, and lower elastic modulus were obtained. PANI was blended with a natural protein, gelatin, and coelectrospun into nanofibers to investigate the potential application of such a blend as conductive scaffolds for tissue engineering purposes. An increasing trend in diameters was reported at increasing PANI concentration. The diameters of nanofibers are important due to the fact that they directly affect conductivities of the mats. There are different reports about the effect of CPs on the diameters of nanofibers. Some of them reported higher diameters; some of them reported lower diameters with increasing CP content. The parameters like the method, concentration, viscosities, types of polymers, and conditions are also important to determine the effect of CPs on the diameters.

8.5 Polyurethane/Polypyrrole Composite Nanofibers

Polyurethane–polypyrrole (PU-PPy) composite nanofibers obtained by using electrospinning were reported. In this study, Py monomer was polymerized in PU matrix by cerium(IV) [ceric ammonium nitrate Ce(IV)] as an oxidant (Fig. 8.8).[234] The effects of the PPy content on the thermal, mechanical, dielectric, and morphological properties of the composites were investigated with different techniques. Morphologies and electrical properties of composite nanofibers were reported.

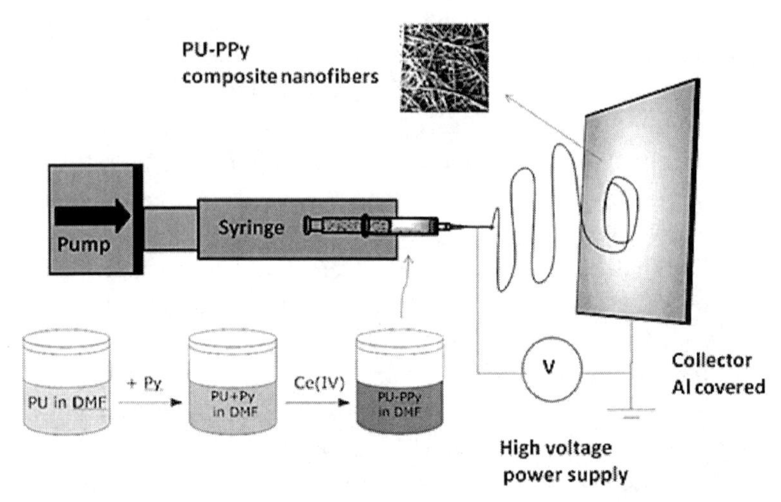

Figure 8.8 Schematic illustration of the composite nanofiber preparation process. Reprinted with permission from Ref. 234, Copyright © 2012, Wiley Periodicals, Inc.

The chemical polymerization of Py by CAN in PU solutions leads to the formation of PU/PPy composites. The composites were characterized by Fourier transform infrared spectrophotometry–attenuated total reflectance (FTIR-ATR), dynamic mechanical analysis (DMA), thermal gravimetric analysis (TGA), differential scanning calorimetry (DSC), X-ray photoelectron spectroscopy (XPS), and SEM measurements. The absorbances of the disordered H-bonded urethane carbonyl decrease with increasing Py concentration. The fraction of the hydrogen-bonded carbonyls is increased and the melting point increases with the increase of PPy content. These indicate the incorporation of PPy into PU may cause the complex due to the intermolecular interaction between the PPy and PU. SEM images of composite nanofibers show good distribution of the second component and the composite solution is proper to form conductive composite nanofibers.

For preparing electrospinning solutions, 2 g PU was dissolved in 5 mL THF and a controlled amount of Py monomers was added to the solution. Finally, CAN was dissolved in ACN (2 mL) and added to the solution.

The electrospinning apparatus consists of a syringe pump, a high-voltage DC power supplier generating a positive DC voltage up to 30 kV DC power supply, and a grounded collector that was covered with aluminum foil. The solution was loaded into a syringe and a positive electrode was clipped onto the syringe needle, having an outer diameter of 0.8 mm. The feeding rate of the polymer solution was controlled by a syringe pump and the solutions were electrospun onto the collector (Fig. 8.9). The syringe pump set at a volume flow rate of 2 mL/h, the applied voltage is 15 kV, the tip-to-collector distance is 10 cm, and the solution concentrations vary with Py concentrations and PU types (for PU nanofibers ~10% w/v; for PU/PPy nanofibers ~20% w/v). A 1:1 (v/v) binary solvent system was used for electrospinning (DMF and THF). All solution preparations and electrospinning were carried out at room temperature.[235]

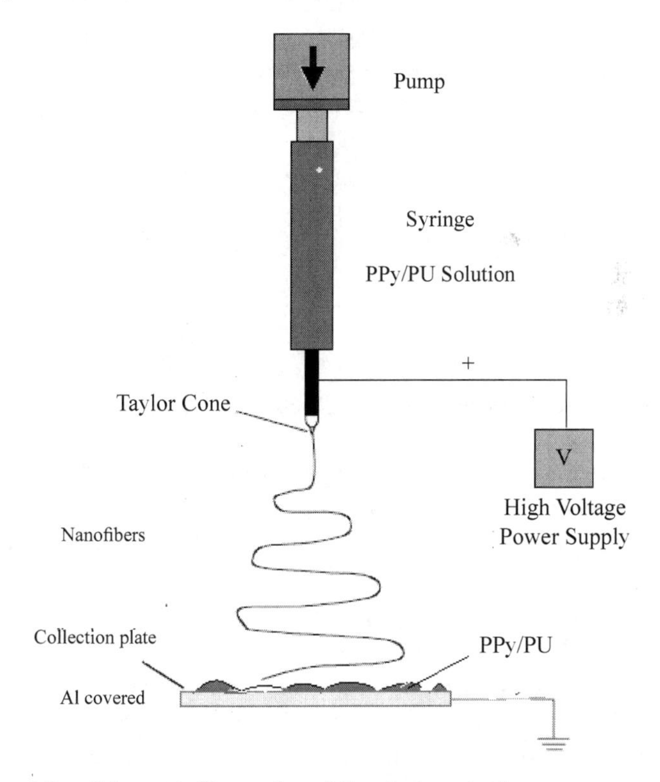

Figure 8.9 Schematic illustration of the electrospinning process.

PUs comprise a polyether, or polyester soft segment, and a diisocyanate-based hard segment. This can be characterized by a two-phase morphology, which is due to the fact that the hard and soft phases are immiscible, leading to the formation of a hard-segment domain, a soft-segment matrix, and an ill-defined interphase.

The domain formation is derived from the intermolecular hydrogen bonding between the hard–hard segments of urethane or urea linkages. The hydrogen bonding is characterized by a frequency shift to the values lower than those corresponding to the free groups (i.e., no hydrogen bonding) in the FTIR-ATR spectrum. Meanwhile, the extent of the frequency shift is usually used as an estimate of hydrogen-bonding strength. Particularly for polyether-based PUs, the fraction of the hydrogen-bonded carbonyls is defined by a hard–hard segment hydrogen bond (N–H· ··O=C bond), which was employed to evaluate the extent of phase separation. On the other hand, the fraction of the hydrogen-bonded ether oxygens (N–H· ··O–) represents the extent of phase mixing between hard and soft segments. The degree of phase separation (DPS) for PU1: 0.35 and PU2: 0.52, and T_g values from DSC are –42°C for PU1 and –24°C for PU2 (Fig. 8.10).[236]

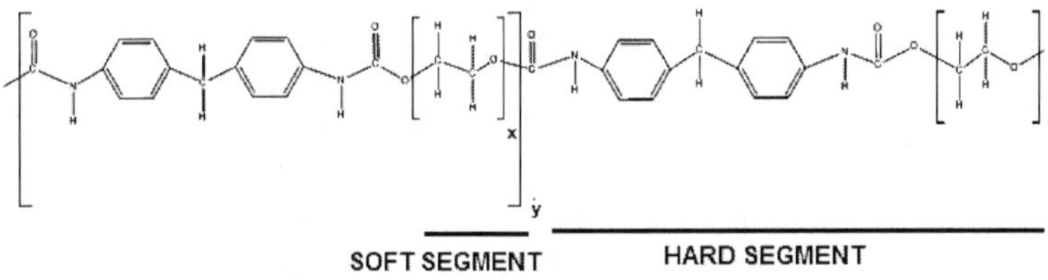

SOFT SEGMENT HARD SEGMENT

Figure 8.10 Polyurethane (for PU1 $X/Y = 0.65$; for PU2 $X/Y = 0.47$).

Figures 8.11 show the FTIR spectra of the carbonyl stretching region ranging from 1600 to 1750 cm^{-1} for PU. The band centered at around 1725 cm^{-1} is attributed to the stretching of free urethane carbonyl groups, whereas the band at 1705 cm^{-1} is assigned to H-bonded urethane carbonyl groups. The absorbances of the free urethane carbonyl groups decrease with increasing PPy content. This is attributable to the fact that the coordination between free urethane carbonyl and PPy which increases with increasing Py concentration, leading to a decrease in the amount of free urethane carbonyl.[236]

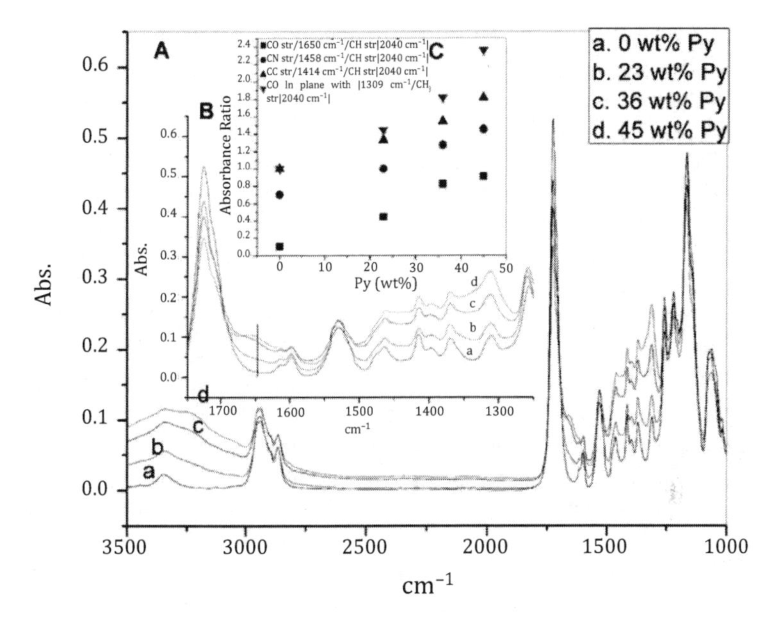

Figure 8.11 (A) FTIR-ATR spectra of the PU and PU–PPy composite films, (B) spectra between 1750 and 1250 cm^{-1}, and (C) absorbance ratios versus Py concentration. Reprinted with permission from Ref. 234, Copyright © 2012, Wiley Periodicals, Inc.

The N–H stretching vibration exhibits an absorption peak at around 3330 cm^{-1} arising from the hydrogen bonding between N–H and carbonyl groups, whereas the free N–H stretching vibration appears at ~3500 cm^{-1}. There exists another shoulder at ~3200 cm^{-1}. This peak corresponds to the NH···O– hydrogen bonding. This result is attributable to the phase-mixed state between hard and soft segments via hydrogen bonding in the polymers and composites, and the shift to a higher frequency with increasing Py content indicates an increase in the H-bonding. This is possibly due to the localization of the electron-rich oxygens on the carbonyl through coordination of the PPy with the H-bonded species.[236]

The absorbances of the urethane carbonyl groups decrease with increasing PPy content and also phase separation becomes less with increasing amounts of PPy.

Thermal properties of PU/PPy composites are investigated using DSC and TGA characterizations. DSC was used to examine the effect of PPy in the matrix. T_g of the composites system increases with increasing Py concentration. This indicates that the PPy chains partially arrested by the PU soft segment.

Figure 8.12 shows DSC results of PU2/PPy composite films with different PPy contents. Different results exist in other composites of different PU. T_g is not influenced by Py as much as the first one. T_g slightly increases with the increasing second component. There is no discernable glass transition in the evaluated temperature range, while sharp endothermic melting peaks were observed for the composites. Those peaks are attributed to the melting temperature of a short-range order within the crystalline structure in the soft-segment domains of the PU chain. The PPy in the PU matrix enhanced the microcrystallization and increased the melting temperature. It was assumed that the introduction of PPy enhanced the phase mixing, hydrogen bonding, and intermolecular packing of microcrystalline domains in the soft segments. PU/PPy composites show a large peak at about 90°C, which is due to the complex. The intensity of the peak increases with the increase of PPy content and the peak position shifts to higher temperatures. In addition, there is also a new exothermic peak at around 60°C in the existence of a high amount of PPy content. Therefore, the incorporation of PPy into PU may cause the complex due to the intermolecular interaction between PPy and PU.

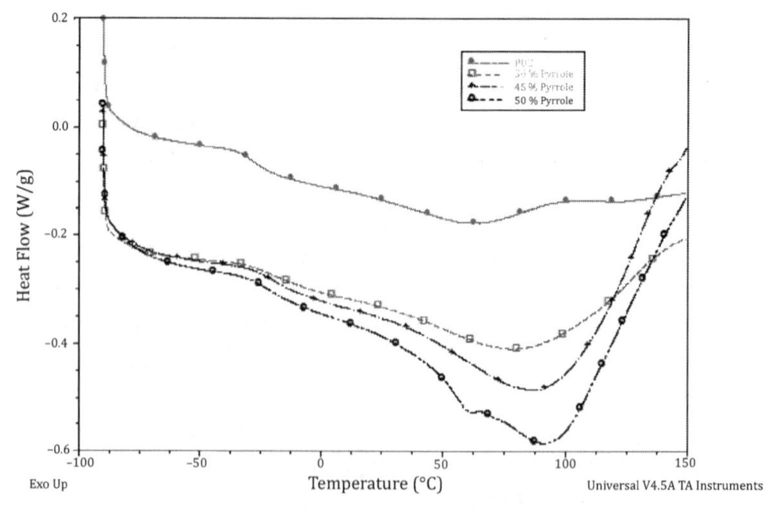

Figure 8.12 DSC thermograms of PU2 and PU2/PPy composites. Reprinted with permission from Ref. 234, Copyright © 2012, Wiley Periodicals, Inc.

TGA thermograms of PU and PU/PPy composites are shown in Fig. 8.13. It is seen that the start of weight loss of pure PU occurs at about 300°C, while PU/ PPy exhibits start of weight losses at about 250°C. This indicates that compared to PU, PPy is easier to be degraded when thermally treated. As is evident from the figure, the TGA thermogram of PU is characterized by a weight loss about when 4% residue was left at about 480°C. The decomposition of the PU/ PPy composite followed a similar tendency as that of pure PU films. Weight loss of the PU/PPy composite started at lower temperatures, which left a residue of about 8% at ~400°C. These results indicate that PU/PPy composite films exhibited good thermal stability like PU films.

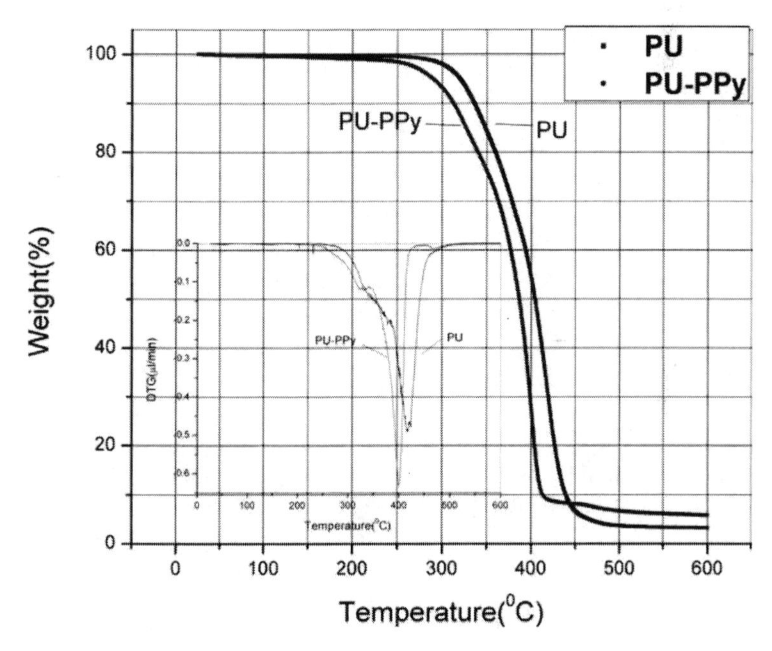

Figure 8.13 TGA thermograms of PU and PU/PPy.

XPS has been used as a quantitative tool to predict the amount of CPs incorporated in polymers. N1s spectra of PPy can be determined. The presence of amine (–NH–) and imine (–N=) components can be identified by the peaks at binding energies of 398.2 and 399.6 eV.

The data revealed that the content of N1s was increased with increasing PPy content, which indicated that PPy chains had been incorporated into PU. It can be seen that the content of C1s decreased with increasing PPy content.

To determine whether PPy content influenced the mechanical properties, mechanical analysis was performed (Fig. 8.14), the results show that after the PPy addition, the mechanical properties change. Increasing the PPy concentration changed the properties of the composite films from elastomeric to ductile plastic-like and finally to brittle.

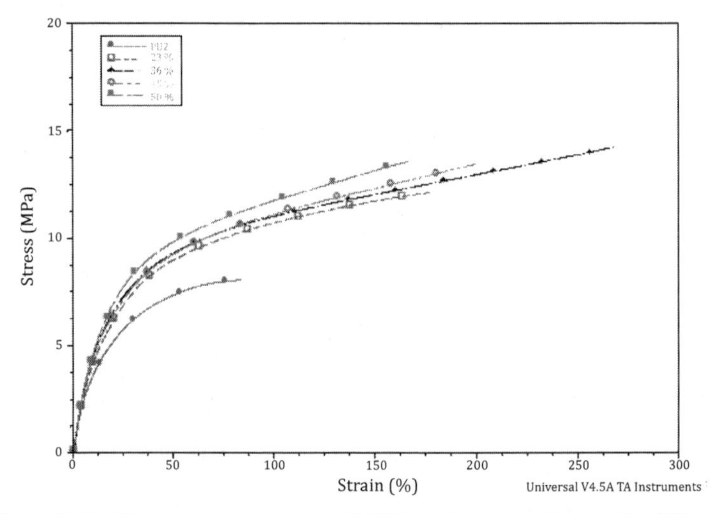

Figure 8.14 Stress–strain curves of PU2 and composites with different amounts of Py. Reprinted with permission from Ref. 234, Copyright © 2012, Wiley Periodicals, Inc.

Figure 8.15 shows SEM images of electrospun PU/PPy composite nanofibers with different PPy content. It is seen that all nanofibers have regular and straight fibrous morphology.

The average fiber diameters decreased with increasing PPy content and the electrospinnability of the solution changed as a result of the interactions between the components in the structure (Fig. 8.16). Due to the strong interaction between PU and PPy, the spinnability of the composite solution was very sensitive to the amount of Py.[242]

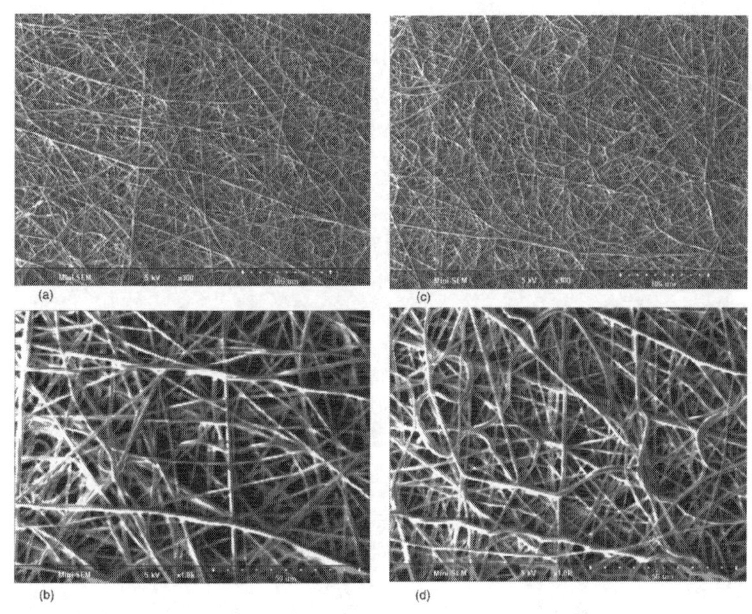

Figure 8.15 Effect of Py content in the PU/PPy fibers: 5% (a), 5% (magnified) (b), 7.5% (c), and 7.5% (magnified) (d). Reprinted with permission from Ref. 234, Copyright © 2012, Wiley Periodicals, Inc.

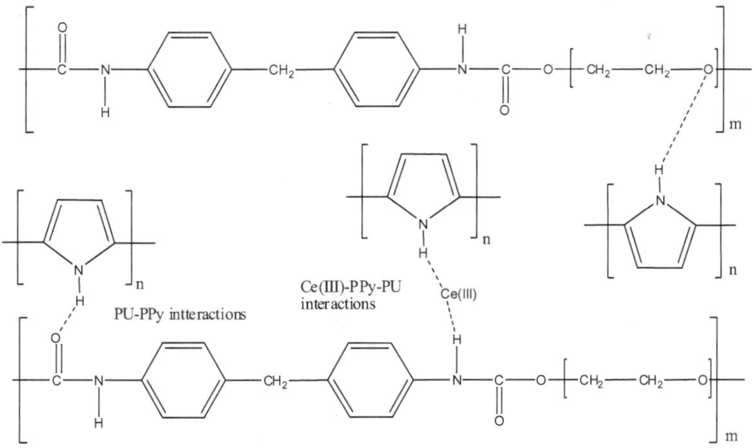

Figure 8.16 Schematic illustrations of the interactions in the composites.

With the increase of PPy content, irregularities, such as so-called "beads on a string" morphology, begin to appear most likely due to the increasing solution conductivity and viscosity induced by the PPy phase (Fig. 8.17).

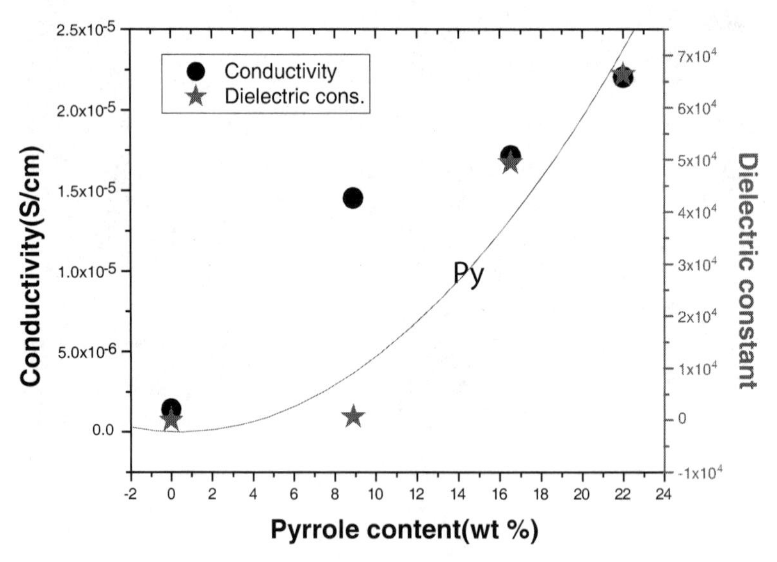

Figure 8.17 Conductivity and dielectric constant vs. increase of Py content. Reprinted with permission from Ref. 234, Copyright © 2012, Wiley Periodicals, Inc.

8.6 Doped Nanofibers and Conductivities

Conductivities of nanofibers were reported in a wide range, depending on different preparation methods, by Chronakis et al.[228] In the first method, PEO was used as the carrier and the conductivities were in the range of 4.9×10^{-8} to 1.2×10^{-5} S/cm, depending on the PPy concentration. In the second and third methods, pure (without carrier) PPy conductive nanofibers were prepared by electrospinning of organic solvent–soluble PPy, using NaDEHS and NaDEHS with PEO to obtain electrospun blends of water-soluble PPy.[1] The electrical conductivity was about 3.5×10^{-4} S/cm for 50 wt.% of [PPy(SO_3H)–DEHS] in the nanofibers (electrospun from a solution with 1.5 wt.% PEO) and the conductivity was 1.1×10^{-4} S/cm for 37.5 wt.% of [PPy(SO_3H)–DEHS] in the nanofibers (electrospun from a solution with 2.5 wt.% PEO), which was nearly 3 orders of magnitude higher than that of the PPy/PEO samples.[1] This difference between the different methods was explained by the higher initial PPy conductivity for the second and the third method. Conductive nonwoven mats composed of PPy were prepared, and the conductivities for electrospun nanowebs were reported about 0.5 S/cm by using the four-probe technique. It was claimed that the fibers had good electrochemical stability for sensor applications. Very high conductivity (14 S/ cm, measured by the four-probe technique) was obtained by using Py deposition and polymerization on salt fibers.[232]

It was reported by Chen et al.[230] that the conductivity was ~1.3×10^{-3} S/cm for Ag/PPy/PAN composite fibrous mats prepared from $AgNO_3$/PAN containing 52% $AgNO_3$. Figure 8.18 shows the conductivities as a function of $AgNO_3$.

An increasing trend in conductivities was reported with increasing CP content in the structure.

PPy–PEO composite nanofibers were prepared by polymerizing Py on PEO fibers by Nair et al.

The sheet conductivities of the PPy–PEO composite nanofiber mats were in the order of 10^{-3} S/cm calculated from the four-probe measurement data.[233] Conductivities of electrospun nanofibers measured by using the four-probe method were about 10^{-5} S/cm. PPy was coated on PS nanofiber mats, and the conductivities of the PS–Cl–PPy and PS–TS–PPy fiber mats were found to be 2×10^{-3} S/ cm and 5×10^{-3} S/cm, respectively. It was demonstrated that the conductivity of the porous fiber mat could be influenced by the amount (PPy/PS ratio), doping, and crystallinity (polymer chain packing) of PPy in the fibers, the void volume, and the connectivity between fibers in the mat. When the PS template of the PS–TS–PPy fiber mat was removed by THF treatment and the electrical conductivity of the remaining material (TS–PPy) was measured, the conductivity increased to 0.13 S/cm by using the four-probe Van der Pauw method.[1]

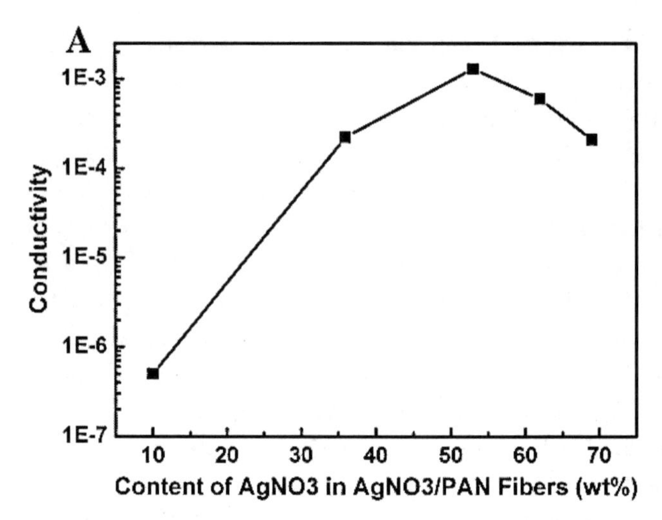

Figure 8.18 Plot of the conductivities (logarithmic scale) of Ag/PPy/PAN fibrous mats vs. the content of $AgNO_3$ in $AgNO_3$/PAN. Reprinted with permission from Ref. 230, Copyright © 2008 Elsevier B.V. All rights reserved.

The conductivities for PANI fibers reported with different dimensions were in agreement with reported results by Cardenas et al. for partially doped PANI and the conductivities were in the range of 10^{-3}–10^{2} S/cm.[214]

Submicron fibers of pure PANI doped with sulfuric acid or hydrochloric acid were prepared and the factors that influence the conductivity were investigated. The doping level and morphology of PANI fibers were the main factors. The higher doping level and more ordered morphology gave a higher conductivity. When the H_2SO_4 concentration increased from 0% to 30%, the doping level increased, and the structural homogeneity improved, so the conductivity increased. If the degree of structural compactness in the fibers reduced, the conductivity decreased.[1]

Conducting electrospun fiber mats based on PLA and PANI blends were reported by Picciani et al.[237] It was observed that the resistivity values decreased with increasing PANI content and increased with increasing fiber diameter. The contact probability among fibers and the formation of the conductive pathways through the sample were introduced as a reason for that result. Thicker fibers had less contact probability in the same mat volume and an increase in fiber diameter results in an increase in void space between fibers. So a decrease in the number of interfiber contact points led to a decrease in conductivity. Changes in crystallinity were also effective.[1]

It was reported by Hong et al.[207] that the volume conductivities increased from 0.5 to 1.5 S/cm as the diffusion time increased from 10 minutes to 4 hours because of the uniform distribution of PANI in the structure of PANI/nylon-6 fiber mats.

As the diffusion time increased, the surface conductivities of the PANI/nylon-6 composite electrospun fiber webs decreased from 0.22 to 0.14 S/cm due to contamination by monomers and oligomers of aniline. Contaminants such as monomers and oligomers of aniline and some alkyl chains served as electrical resistants.[1]

PANI nanoparticles were doped with DBSA and electrospun with nylon-6 by Hong et al.[220] Conductivities of different forms were compared. The electrical conductivity of PANI (DBSA) particle pellets was about 4.27×10^{-2} S/cm, the conductivity of the PANI (DBSA)–nylon-6 film was about 1.68×10^{-4} S/cm, and the conductivity of the PANI (DBSA)–nylon 6 electrospun fiber web was about 6.19×10^{-7} S/cm. It was concluded that when the PANI (DBSA–nylon 6) composite solution was electrospun by electric power, the overall crystallinity of the composite polymer decreased so the conductivity decreased. This was explained with the rapid evaporation of the solvent during the electrospinning process.

Conducting PANI/PLA nanofibers by electrospinning were reported and conductivities between bulk and nanofiber films were compared.[238] It was found that nanofiber mats had lower crystallinity due to the fact that rapid evaporation of the solvent prevents chains to organize into a suitable crystal structure. The high porosity of the nonwoven mats and lower crystallinity resulted in a decrease in the electrical conductivity.

Ultrafine fibers of PANI doped with CSA were blended with PEO by electrospinning.[239] By controlling the ratio of PANI to PEO, desired conductivities were obtained. The comparison between cast films and nanofiber mats was reported. High porosity of nanofiber mats leads to lower conductivities compared to cast films but nanofiber mats have advantages like quick dedoping due to a higher surface area.

PANI/silica hybrid nanofiber webs were prepared by using two different methods.[215] The first one was in situ polymerization of aniline-doped silica nanofibers in an ammonium persulfate solution and the second one was immersing the silica nanofibers into a PANI solution. Effects of different polymerization parameters on conductivity were investigated. The PANI/silica hybrid nanofiber web electric conductivity increased with increasing monomer concentration. The electrical conductivities of the hybrid web were 5×10^{-5}, 1.7×10^{-3}, 4.5×10^{-3}, 3.2×10^{-3}, and 1.07 S/cm for 0.2, 0.5, 0.7, and 1.0 M aniline solutions and pure polyaniline, respectively. The electrical conductivity showed the maximum at 1.0 of the molar ratio of the oxidant and aniline and decreased with an increase of the oxidant concentration. This result was explained by the prevention of polymerization by excess oxidant. The molar ratio of oxidant and aniline is generally about 1.0 for synthesis of PANI.

For the hybrid web prepared by the first method, the electrical conductivity and amount of coated PANI increased with the increase of the aniline concentration. The electrical conductivity decreased with an increase of the number average degree of polymerizations. The electrical conductivity increased slightly with an increase of the polymerization time until 30 minutes. The experimental results indicated that the desirable preparation conditions of the PANI/silica nanofiber hybrid web by the in situ polymerization were 0.7 M aniline solution, and 10–30 minutes of polymerization time, 1.0 of the molar ratio of oxidant and aniline, and 0.1–0.5 M of the dopant concentration.[1]

P3HT/PEO blend nanofibers were produced by Laforgue et al.[96] The maximum electrical conductivity for unaligned mats was 0.16 S/cm and increased to 0.3 S/cm when the nanofibers were aligned. PU–PPy nanofibers were prepared by polymerizing Py monomer in the PU matrix by means of oxidative polymerization with cerium (IV) in DMF by Yanilmaz et al.[234] The comparison between PU nanofibers and PU–PPy composite nanofibers were reported. The AC conductivities of the PU nanofibers without Py and with 5% Py were about 7×10^{-7} and 1.4×10^{-6} S/cm, respectively, at 10^{7} Hz. The composite nanofibers exhibited a high dielectric constant and tan δ values in the low- and radio-frequency ranges, and they could be used in charge-storing devices, decoupling capacitors, and electromagnetic interference (EMI) applications[1] (Fig. 8.19).

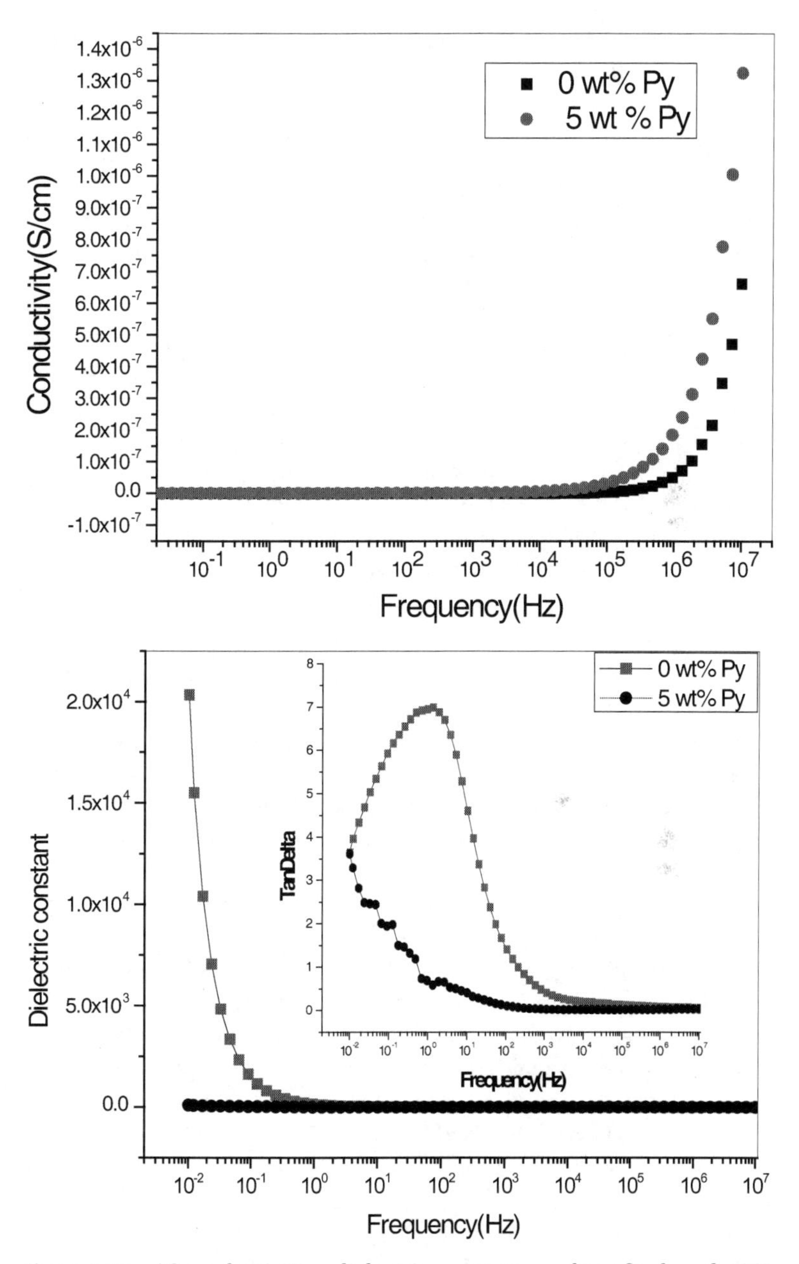

Figure 8.19 AC conductivities, dielectric constants, and tan δ values for PU and PU–PPy nanofibers. Reprinted with permission from Ref. 234, Copyright © 2012, Wiley Periodicals, Inc.

High porosities and lower crystallinities of the nanofiber structure are the disadvantages of nanofiber mats for high-conductivity applications. On the other hand a high specific surface area improves performances for many applications. Conductivities can be affected by several factors, including types of polymers and other chemicals (solvents, dopants, oxidizing agents, etc.), ratios of the components, methods, and ambient parameters.[1]

8.7 Light-Emitting Polymer Nanofibers

Light-emitting 1D nanomaterials made by conjugated polymers have unique features because they combine optoelectronic properties of semiconductors with structural properties of polymers. Organic nanowires, nanofibers, and nanotubes can be produced by different techniques, including template-assisted synthesis and vacuum sublimation. However, the low throughput of many of these approaches limits their application in photonic integrated systems.

Light-emitting polymer nanofibers can be produced by electrospinning, a high-throughput method, to realize 1D structures with typical diameters down to tens of nanometers. Electrospinning stretches a viscous polymer solution along one axis by applying an electrostatic field between a metallic needle and a collector. Applying the method to conjugated polymers is particularly challenging due to the limited solubility and relatively poor viscoelasticity of these compounds. However, some of the challenges can be partially overcome by making the final solution more suitable for processing. Light-emitting nanofibers can be fabricated with different conjugated polymers, copolymers, and their blends, displaying emissions tunable in the whole visible range. Figure 8.20 shows typical photoluminescence confocal and scanning electron micrographs of conjugated polymer fibers, exhibiting average diameters in the range of 100 nm–10 μm, depending on the processing parameters (applied voltage, solution feeding rate, and needle–collector distance) and type of tetraakylammonium halide (Fig. 8.20). Emission can be further tuned by Förster energy transfer, that is, by means of the dipole–dipole nonradiative energy transfer of the excitation from a donor molecule to a suitably chosen acceptor molecule. In particular, white light emission can be achieved.*

---

*Excerpt reprinted with permission from Ref. 240, Copyright 2011, SPIE.

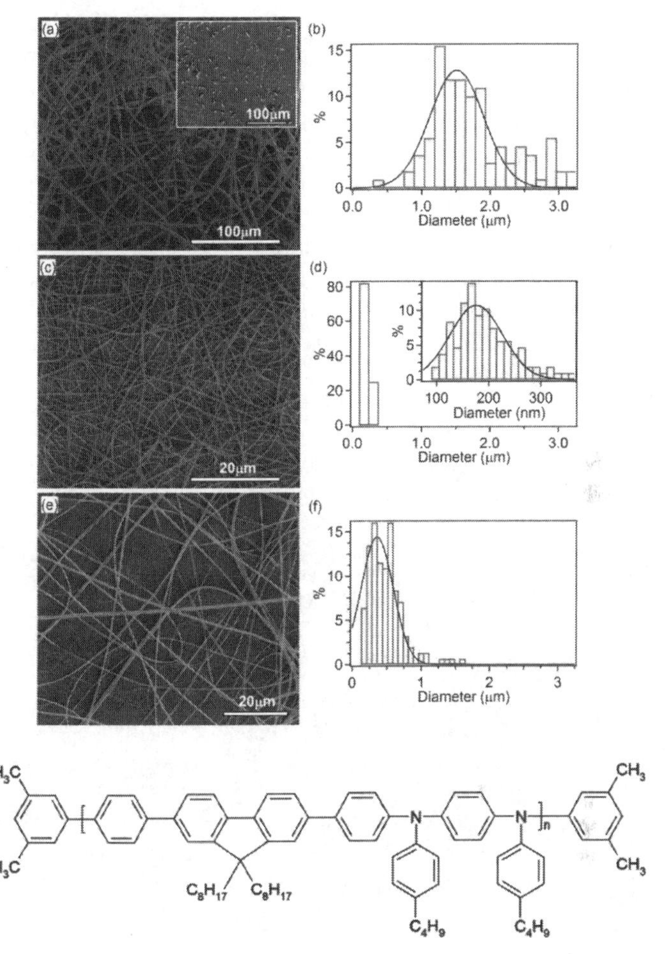

Figure 8.20 (a–b) SEM micrograph and nanofiber distribution of light-emitting electrospun nanofibers of poly[(9,9-dioctylfluorenyl-2,7-diyl)-co-(*N*,*N*-diphenyl)-*N*,*N*-di(*p*-butyl-oxy-phenyl)-1,4-diaminobenzene)] (PFO–PBAB) in THF/DMF, (c–d) in $CHCl_3$/TBAI, and (e–f) $CHCl_3$/TBAB. Reprinted from Ref. 241, Copyright 2013, American Chemical Society. http://pubs.acs.org/doi/pdf/10.1021/ma400145a

8.7.1 Light Emission and Waveguiding in Conjugated Polymer Nanofibers Electrospun from Organic Salt–Added Solutions

Light-emitting electrospun nanofibers of poly[(9,9-dioctylfluorenyl-2,7-diyl)-co-(*N*,*N*′-diphenyl)-*N*,*N*′-di(*p*-butyl-oxy-phenyl)-1,4-diaminobenzene)] (PFO–PBAB) are produced by electrospinning under different experimental conditions (Fig. 8.21).[241] In particular, uniform fibers with an average diameter of 180 nm are obtained by adding an organic salt to the electrospinning solution. The spectroscopic assessment reveals that the presence of the organic salt does not change the optical properties of the active material, and thus provides an alternative approach for the fabrication of highly emissive conjugated polymer nanofibers. The produced nanofibers display self-waveguiding of light, and polarized photoluminescence, which is especially promising for embedding active electrospun fibers in sensing and nanophotonic devices.[242]

The optical properties of the nanofibers can also be tailored by including a periodic, wavelength-scale microstructure in the fibers. This can be achieved through specifically developed lithographic techniques such as room-temperature nanoimprint lithography (RT-NIL).*

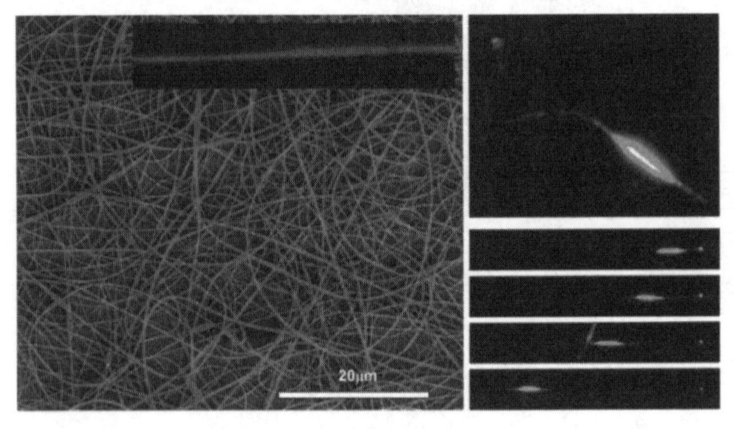

Figure 8.21 Light-emitting electrospun nanofibers of PFO–PBAB. Reprinted from Ref. 241, Copyright 2013, American Chemical Society. http://pubs.acs.org/doi/pdf/10.1021/ma400145a

This technology is based on the deformation, under applied pressure, of a polymeric material by a nanostructured silicon template in contact with the target polymer surface. Being performed at room temperature, the process preserves the emission properties of the conjugated materials. In fact, these compounds are usually very sensitive to degradation effects caused by elevated temperatures or by the interaction with high-energy light (UV) or charged particles, which

*Excerpt reprinted with permission from Ref. 243, Copyright 2010, IOP Publishing, and Ref. 244, Copyright 2008, Macmillan Publishers Limited.

are used in most lithographic techniques. By RT-NIL, authors obtained 1D gratings with periods between 500 nm and 650 nm on light-emitting fibers, thereafter showing an increased forward emission due to Bragg scattering of waveguided modes.*

By the control of the polarization of the emission by choosing the grating orientation with respect to the fiber axis, and high stretching of the polymer solution during the electrospinning process, the polymer backbone aligns with the fiber axis. The emission polarized along the same axis is enhanced as a result. By imprinting a 1D grating parallel or perpendicular to the fiber axis, the polarization of the emission along the fiber axis can be enhanced or depressed, respectively.*

The possibility of fabricating organic semiconductor, light-emitting nanofibers by a simple, low-cost, and high-throughput technique may open new perspectives in plastic photonics. It was exploited such building blocks within lasers and field-effect transistors and as miniature, polarized light sources in prototype lab-on-a-chip devices, which can be further developed to apply photonic integrated circuits where fibers can act as light sources, waveguides, and detectors.*

8.8 Bioapplications

Due to the inherent conductivity and electroactivity of conducting polymers, they act as suitable substrates for the in vitro study of excitable cells, including skeletal muscle cells. Biocompatible conducting polymers, such as polypyrrole, polyaniline, and PEDOT, have been used as substrates for the culture of a range of cell types, including PC12 cells, endothelial cells, fibroblasts, and keratinocytes.

To optimize the use of the aligned fibers as a cell growth platform for in vitro muscle tissue development, the alignment of differentiated myotubes into parallel muscle fibers on these structures was investigated. Primary myoblasts were seeded in cell proliferation media onto gold mylar coated with the aligned fibers at a range of fiber densities. The fiber densities ranged from high density to low density (100–300 micrometer separation of fibers, with exposed gold mylar substrate). The cells did not adhere well to areas of high density linearly aligned fibers, that is, where there was no separation between the fibers. When the fibers were separated by intervening areas of gold mylar, at medium and low densities the myoblasts adhered to and proliferated on the fibers

*Excerpt reprinted with permission from Ref. 241,Copyright 2011, SPIE.

and on the gold mylar between the fibers.[246] When the polymer fibers were in the medium density range (15–100 micrometer separation), differentiated primary myotubes tended to align with both the poly(HET) and poly(OTE) fibers [electrospun fibers from an ester-functionalized organic solvent-soluble polythiophene (poly-octanoic acid 2-thiophen-3-yl-ethyl ester)] [Fig. 8.22(A–C)]. At lower fiber densities [Fig. 8.22(E,F)], there was no significant alignment of myotubes with the fibers and, as was the case for polymer films [Fig. 8.22(D)], the myotubes were randomly oriented and branched. Thus, electrospun polymer fibers, aligned with separations of 15–100 micrometer, guided the differentiation of myoblasts into aligned, linear myotubes, a desirable outcome for in vitro skeletal muscle tissue engineering applications.[246]

CPs such as PPy, PANI, polythiophene (PT), and PEDOT show biocompatibility, conductivity, reversible oxidation, redox stability, and excellent electrical and optical properties. These make them suitable for hydrophobicity for cell adhesion, desired for tissue engineering applications.[245] As an example, PPy-coated electrospun poly(lactic-co-glycolic acid) (PLGA) nanofibers (PPy–PLGA) for neural tissue applications were studied. It was reported that these nanofibers supported the growth and differentiation of rat pheochromocytoma 12 (PC12) cells and hippocampal neurons. Electrical stimulation of neurons on electroconducting scaffolds was also shown to demonstrate the use of PPy–PLGA meshes as potential nerve tissue engineering scaffolds delivering electrical cues through nanofibers. PC12 cells on PPy–RF (random) and PPy–AF (aligned) fibers at the potentials of 10 mV/cm and 100 mV/cm were electrically stimulated and neurite lengths, percentages of neurite-bearing cells, and numbers of neurites per cell were analyzed.[245] As a result, more neurite-bearing PC12 cells and longer neurites were observed with electrical stimulation compared to unstimulated controls. It was concluded that PPy–PLGA meshes were appropriate for neuronal applications and showed topographies for modulating cellular interactions comparable to the PLGA control nanofibers.

The thiophene family of conducting polymers offers unique flexibility for tailoring of polymer properties as a result of the ease of functionalization of the parent monomer, i.e., an ester-functionalized organic solvent-soluble *polythiophene* (*poly-octanoic acid 2-thiophen-3-yl-ethyl ester*) changing the properties of PTh and resulting from post-polymerization hydrolysis of the ester linkage. The polymer films supported the proliferation and differentiation of both primary and transformed skeletal muscle myoblasts. In addition, aligned electrospun fibers formed from the polymers provided scaffolds for the guided differentiation of linearly aligned primary myotubes, suggesting their suitability as three-dimensional substrates for the in vitro engineering of skeletal muscle tissue (Fig 8.22).*

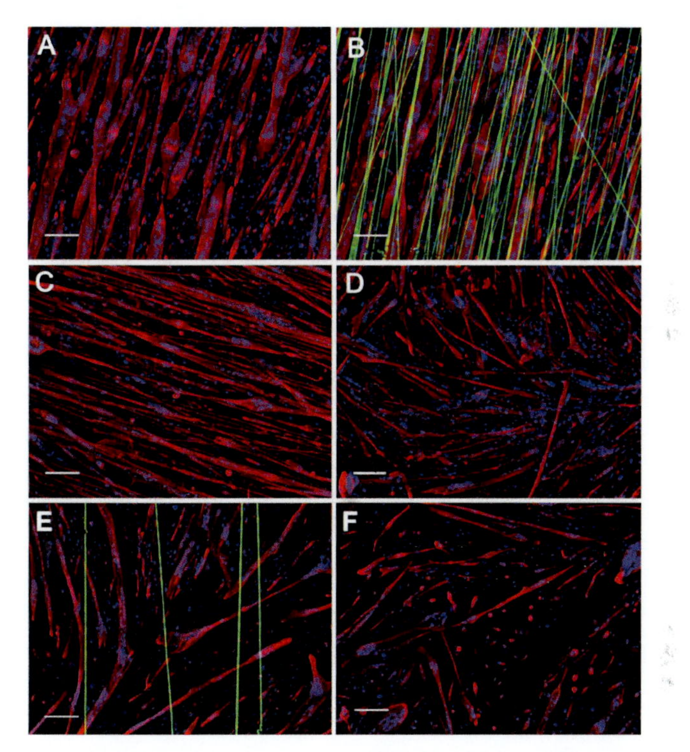

Figure 8.22 Fluorescence images of differentiated primary myotubes, aligned along medium density fibers of poly(OTE) (A and B) or poly(HET) (C). A and B are the same image without (A) and with (B) fibers shown. Myotubes differentiated on the gold mylar substrate (D) or on low densities of poly(OTE) (E) or poly(HET) (F) fibers were randomly oriented and branched. Myotube nuclei were stained with DAPI (blue) whereas the myoblast protein desmin was immunostained with an anti-Desmin primary antibody and Alexa 546-tagged (red) secondary antibody. Scale bars represent 100 micrometer. Reprinted with permission from Ref. 246, Copyright © 2013, American Chemical Society.

*Excerpt reprinted with permission from Ref. 246, Copyright 2013, American Chemical Society.

8.8.1 Polymer Nanofibers for Biomedical, Biotechnological and Capacitor Applications

Polymer nanofibers for biomedical and biotechnological applications such as tissue engineering, controlled drug release, wound dressing, medical implants, nanocomposites for dental restoration, molecular separation, biosensors, and preservation of bioactive agents were reviewed.[247]

CPs in different forms like nanofibers and thin films for tissue engineering applications were evaluated by Bendrea et al.[248] Conducting C-PANI was blended with poly(L-lactide-co-ε-caprolactone) (PLCL) and then electrospun to prepare a uniform nanofiber scaffold. This scaffold combined the elastic properties (comes from the PLCL domain) with electrical activity (due to conducting C-PANI) at the nanometer-scale features. It was proved that a nanoscale structure with PANI led to a high pore volume, interconnective pores, a uniform mean fiber diameter, and significantly increased conductivity.[1]

CPs have also been considered as potential materials for sensors because of their inherent optical, electronic, and mechanical transduction mechanisms. These sensors have advantages such as relative low cost, reversible signal transduction, high sensitivities, and rapid response at room temperature. PEDOT-PSS/poly(*N*-vinyl pyrrolidone) (PVP) composite nanofibers were prepared by the blending technique and electrospinning. PEDOT-PSS/PVP nanofibers showed good reversibility and reproducibility in organic vapor sensing.[249] Compared with PVP nanofibers, PEDOT-PSS/PVP nanofibers exhibited better organic vapor-sensing performances to ethanol, methanol, THF, and acetone.

Electrospun nanofibers have been confirmed to be good candidates for ultrasensitive gas sensors due to the improved surface-area-to-volume ratios of coatings.[250] A higher surface area led to higher sensitivity and fast response time.

Preparation of PANI nanofiber humidity sensors produced by electrospinning from the DMF solution of PANI, poly(vinyl butyral) (PVB), and PEO were shown.[251] It was concluded that PANI nanofibers with some beads and a small content of PEO revealed high sensitivity, fast response, and small hysteresis because beads could help to improve adhesion to the electrode (that enhance electrical contact and sensing ability) and PEO helped to increase the hydrophilicity of the PANI nanofibers and humidity responses.

An electrospinning solution was prepared as follows: 0.1 g of PANI/PSSA was dissolved in 5 mL DMF and a different amount of PEO, and PVB was used as a matrix (impedance change from 10^7 to 10^3 Ω over the range of 11–95% relative humidity).[1]

CPs have been studied to apply for the electrodes of chargeable batteries, fuel cells, and electrochemical capacitors because of the fact that they have high electrical conductivity, light weight, and easy processibility, which makes them suitable for electrodes. It was reported that electrospun PPy/sulfonated poly(styrene ethylene butylstyrene) fibers may enhance electrochemical capacity due to their high doping level and the ease of charge transfers reactions because of high conductivity. In the study, a PPy composite nanofiber electrode by electrospinning was compared with the electrode by a casting method. The electrospun nanofibers showed higher charge/discharge specific capacity than the granular type by the casting method. This result was explained with the reduction of interfacial resistance caused by the decrease of the contact area. In another study, nanostructured PANI has been tested for sensor, actuator, supercapacitor, and gas separation membrane applications. For the preparation of hollow PANI nanofibers, in situ polymerization of aniline was carried out at the temperature of 0–5 °C. Prepared electrospun PAA fiber membrane was treated with ammonium persulfate (APS)/HCl to initiate the polymerization. Different aniline concentrations (0.01, 0.03, and 0.05 M) were used for polymerization over 12 h, respectively, with the molar ratio of aniline/APS fixed at 2/1. The as-prepared PAA/PANI core/shell nanofibers were washed with dimethylacetamide (DMAc) to remove the PAA core. Thus, hollow PANI nanofibers with different diameters (400, 450, 550 nm respectively) were obtained and correspondingly marked as H-PANI 1#, 2#, and 3# for the three aniline concentrations. PANI powder was in situ polymerized under the same condition but without the template and marked as P-PANI for comparative purposes (Fig. 8.23). H-PANI 1# with thinner wall thickness exhibits higher capacitance of 601 F g^{-1} at 1 A g^{-1}, which can be attributed to the shortened charge transport distance of the electroactive ions into the far inner layer of PANI.[252]

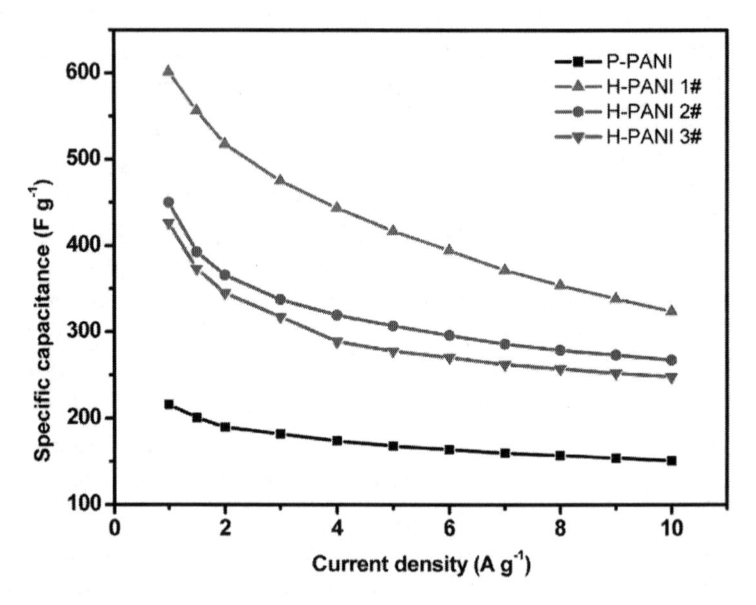

Figure 8.23 Specific capacitance plots of P-PANI, H-PANI 1#, H-PANI 2#, and H-PANI 3# at different current densities, respectively. Reprinted with permission from Ref. 252, Copyright © 2013, American Chemical Society.

8.9 Concluding Remarks

The introduction of conducting materials and especially CPs into nanofiber mats has the potential to provide sufficient conductivity for different applications. Controllable conductivity of these materials is also an advantage for different purposes. In this book, the preparation and properties of semi- or conductive nanofibers in the presence of CPs by using the electrospinning technique are reviewed and presented. The challenges and limitations of different mediums are reported and many potential applications are mentioned. The main requirements for many applications are improved conductivity and uniform morphology. These can be improved by maximizing CPs without affecting morphology. Conductivities can be affected by several factors, including types of polymers and other chemicals (solvents, dopants, oxidizing agents, etc.), solvent properties, ratios of the components, methods, and ambient parameters. The selection of polymers, solvents, and doping and oxidizing agents is important to obtain required properties, including uniform morphology and high conductivity. High porosities and lower crystallinities are the disadvantages of nanofiber mats for high-conductivity applications. On the other hand, a higher specific surface area due to the nanofiber structure improves performances for different applications.

Bibliography

1. Yanilmaz M. and Sarac A. S., A review: Effect of conductive polymers on the conductivities of electrospun mats, *Text. Res. J.*, 2014, 84, 12, 1325–1342.
2. Jian F., HaiTao N., Tong L., and XunGai W., Applications of electrospun nanofibers, *Chin. Sci. Bull.,* 2008, 53, 15, 2265–2286.
3. Nayak R., Padhye R., Kyratzis I. L., Truong Y. B., and Arnold L., Recent advances in nanofibre fabrication techniques, *Text. Res. J.,* 2011, 82, 2, 129–147.
4. Miao J., Miyauchi M., Simmons T. R., Dordick J. S., and Linhardt R. J., Electrospinning of nanomaterials and applications in electronic components and devices, *J. Nanosci. Technol.,* 2010, 10, 5507–5519.
5. Li X., Hao X., Yu H., and Na H., Fabrication of polyacrylonitrile/ polypyrrole (PAN/Ppy) composite nanofibres and nanospheres with core–shell structures by electrospinning, *Mater. Lett.,* 2008, 62, 1155–1158.
6. Hussain D., Loyal F., Greiner A., and Wendorff J. H., Structure property correlations for electrospun nanofiber nonwovens, *Polymer,* 2010, 51, 3989–3997.
7. Subbiah T., Bhat G. S., Tock R. W., Parameswaran S., and Ramkumar S. S., Electrospinning of nanofibers, *J. Appl. Polym. Sci.,* 2005, 96, 557–569.
8. Baji A., Mai Y. W., Wong S. C., Abtahi M., and Chen P., Electrospinning of polymer nanofibers: effects on oriented morphology, structures and tensile properties, *Compos. Sci. Technol.,* 2010, 70, 703–718.
9. Huang Z. M., Zhang Y. Z., Kotakic M., and Ramakrishna S., A review on polymer nanofibers by electrospinning and their applications in nanocomposites, *Compos. Sci. Technol.,* 2003, 63, 2223–2253.
10. Greiner A. and Wendorff J. H., Functional self-assembled nanofibers by electrospinning, *Adv. Polym. Sci.,* 2008, 219, 107–171.

11. Cavaliere S., Subianto S., Savych I., Jones D. J., and Rozière J., Electrospinning: designed architectures for energy conversion and storage devices, *Energy Environ. Sci.,* 2011, 4, 4761–4785.
12. Wu H., Pan W., Lin D., and Li H., Electrospinning of ceramic nanofibers: Fabrication, assembly and applications, *J. Adv. Ceram.,* 2012, 1, 1, 2–23, DOI: 10.1007/s40145-012-0002-4.
13. Haghi A. K., *Electrospinning of Nanofibers in Textiles*, Apple Academic Press, NJ, 2012.
14. Venugopal J. and Ramakrishna S., Applications of polymer nanofibers in biomedicine and biotechnology, *Applied Biochemistry and Biotechnology* 2005, 125, 147.
15. Frenot A. and Chronakis I. S., Polymer nanofibers assembled by electrospinning, *Curr. Opin. Colloid Interface Sci.*, 2003, 8, 64–75.
16. Electrospinning http://www.electro-spinning.com/
17. Electrospinning. Retrieved from http://en.wikipedia.org/wiki/Electrospinning
18. Chase G. G., Sunthorn Varabhas J., and Reneker D. H., New methods to electrospin nanofibers, *Journal of Engineered Fibers and Fabrics* http://www.jeffjournal.org 2011, 6, 3, 32–38.
19. Wilkes, G., Electrospinning, http://www.che.vt.edu//Faculty/Wilkes/GLW/electrospinning/electrspinning.html.
20. Nanofiber. Retrieved from https://en.wikipedia.org/w/index.php?title=Nanofiber&oldid=708560966
21. Son W. K., Youk J. H., Lee T. S., and Park W. H., The effects of solution properties and polyelectrolyte on electrospinning of ultrafme poly(ethylene oxide) fibers, *Polymer*, 2004, 45, 2959–2966.
22. Lee K. H., Kim H. Y., Ra Y. M., and Lee D. R., Characterization of nanostructured poly(e-caprolactone) nonwoven mats via electrospinning, *Polymer,* 2003, 44, 1287–1294.
23. Hsu C. M. and Shivakumar S. N., N-dimethylformamide additions to the solution for the electrospinning of poly(e-caprolactone) nanofibers, *Macromol. Mater. Eng.*, 2004, 289, 334–340.
24. Avci M. Z. and Sarac A. S., Transparent poly(methyl methacrylate-co-butyl acrylate) nanofibers, *J. Appl. Polym. Sci.*, 2013, 130, 4264–4272 DOI:10.1002/APP.39705.
25. Buchko C. J., Chen L. C., Shen Y., and Martin D. C., Processing and microstructural characterization of porous biocompatible protein polymer thin films, *Polymer*, 1999, 40, 7397–7407.

26. Megelski S., Stephens J. S., Chase D. B., and Rabolt J. F., Micro- and nanostructured surface morphology on electrospun polymer fibers, *Macromolecules*, 2002, 35, 8456–8466f.

27. Taylor G., Disintegration of water drops in an electric field, *Proc. R. Soc. Lond.*, 1964, 280, 383–397.

28. Zhong X. H., Kim K. S., Fang D. F., Ran S. F., Hsiao B. S., and Chu B., Structure and process relationship of electrospun bioabsorbable nanofiber membranes, *Polymer*, 2002, 43, 4403–4412.

29. Deitzel J. M., Kosik W., McKnight S. H., Tan N. C. B., DeSimone J. M., and Crette S., Electrospinning of polymer nanofibers with specific surface chemistry, *Polymer*, 2002, 43, 1025–1029.

30. Yuan X., Zhang Y., Dong C., and Sheng J., Morphology of ultrafine polysulfone fibers prepared by electrospinning, *Polym. Int.*, 2004, 53, 1704–1710.

31. Jarusuwannapoom T., Hongrojjanawiwat W., Jitjaicham S., Wannatong L., NithitanakulM., Pattamaprom C., Koombhongse P., Rangkupan R., and Supaphol P., Effect of solvents on electro-spinnability of polystyrene solutions and morphological appearance of resulting electrospun polystyrene fibers, *Eur. Polym. J.*, 2005, 41, 409–421.

32. Hager B. L. and Berry G. C., Moderately concentrated solutions of polystyrene. I. Viscosity as a function of concentration, temperature, and molecular weight. *J. Polym. Sci., Polym. Phys. Ed.*, 1982, 20, 911–928.

33. https://theses.lib.vt.edu/theses/available/etd-12142004-090652/unrestricted/CHAPTER-2.pdf

34. Gupta P., Elkins C., Long T. E., and Wilkes G. L., Electrospinning of linear homopolymers of poly(methyl methacrylate): exploring relationships between fiber formation, viscosity, molecular weight and concentration in a good solvent, *Polymer*, 2005, 46, I 13, 4799–4810.

35. Basu S., Gogoi N., Sharma S., Jassal M., and Agrawal A. K., Role of elasticity in control of diameter of electrospun pan nanofibers, *Fibre Polym.*, 2013, 14(6), 950–956, DOI:10.1007/s12221-013-0950-5.

36. Shin Y. M., Hohman M. M., Brenner M. P., and Rutledge G. C., Experimental characterization of electrospinning: the electrically forced jet and instabilities, *Polymer*, 2001, 42, 9955–9967.

37. Fridrikh S. V., Yu J. H., Brenner M. P., and Rutledge G. C., Controlling the fiber diameter during electrospinning, *Phys. Rev. Lett.*, 2003, 90, 144502:1–144502:4.

38. Zander N. E., Hierarchically structured electrospun fibers. *Polymer*, 2013, 5(1), 19–44.

39. Satici M. T. and Sarac A. S., Synthesis and characterization of poly(acrylonitrile-co-vinylacetate)/Fe2O3@PEDOT core-shell nanocapsules and nanofibers, *International Journal of Polymeric Materials and Polymeric Biomaterials*, 2015, 64(11), 597–609.

40. Carroll C. P. and Joo Y. L., Axisymmetric instabilities in electrospinning of highly conducting, viscoelastic polymer solutions, *Phys. Fluids*, 2009, 21, 103101; http://dx.doi.org/10.1063/1.3246024

41. Sen S. and Sarac A. S., Electrospun nanofibers of acrylonitrile and itaconic acid copolymers and their stabilization, *J. Adv. Chem.*, 2014, 6, 2958–2981.

42. Ohkawa K., Cha D., Kim H., Nishida A., and Yamamoto H., Electrospinning of chitosan, *Macromol. Rapid Commun.*, 2004, 25, 1600–1605.

43. Sheeny K., Levengood L., and Zhang M., Chitosan-based scaffolds for bone tissue engineering (feature article), *J. Mater. Chem. B*, 2014, 2, 3161–3184, DOI:10.1039/C4TB00027G.

44. Yu D. G., Williams G. R., Yang J. H., Wang X., Qian W., et al., Chitosan Nanoparticles self-assembled from electrospun composite nanofibers, *J. Textile Sci. Engg.*, 2012, 2:107, DOI: 10.4172/2165–8064. 1000107.

45. Li F., Zhao Y., and Song Y., 2010. *Core-Shell Nanofibers: Nano Channel and Capsule by Coaxial Electrospinning, Nanofibers*, Ashok Kumar (Ed.), InTech. ISBN: 978-953-7619-86-2, DOI:10.5772/8166.

46. Eliton S. M., Gregory M. Gl., Artur P. K.,William J. O., and Luiz H. C. M., Solution blow spinning: A new method to produce micro- and nanofibers from polymer solutions, *J. Appl. Polym. Sci.*, 2009, 113, 4, 2322–2330, DOI: 10.1002/app. 30275.

47. Yerlikaya Y., Unsal C., and Sarac A. S., Nanofibers of poly(acrylonitrile-co-methylacrylate)/polypyrrole core-shell nanoparticles, *Adv. Sci. Eng. Med.*, 2014, 6, 301–310.

48. Unsal C., Kalaoglu F., Karakas H., and Sarac A. S., Polypyrrole/poly(acrylonitrile-co-butyl acrylate), *Comp. Adv. Polym. Technol.*, 2013, 32, S1, E784–E792, DOI:10.1002/adv.21321.

49. Guler Z., Erkoc P., and Sarac A. S, Electrochemical impedance spectroscopic study of single-stranded DNA-immobilized electroactive polypyrrole-coated electrospun poly(ε-caprolactone) nanofibers, *Materials Express* 2015, 5(4), 269–279.

50. Golshaei R., Guler Z., Ünsal C., and Sarac A. S., In situ spectroscopic and electrochemical impedance study of gold/poly (anthranilic acid) core/shell nanoparticles, *Eur. Polym. J.*, 2015, 66, 502–512.

51. Fang J., Niu H. T., Lin T., Wang X., 2008, Applications of electrospun nanofibers, *Chinese Science Bulletin,* 2008, 53, 2265–2286.

52. Chronakis I. S., Novel nanocomposites and nanoceramics based on polymer nanofibers using electrospinning process: a review, *J. Mater. Proc. Technol.,* 2005, 167, 283–293.

53. Bergshoef M. M. and Vancso G. J., Transparent nanocomposites with ultrathin, electrospun nylon-4, 6 fiber reinforcement, *Adv. Mater.,* 1999, 11, 1362–1365.

54. Lee D. Y., Lee K.-H., Kim B.-Y., and Cho N.-I., Silver nanoparticles dispersed in electrospun polyacrylonitrile nanofibers via chemical reduction, *J. Sol-Gel Sci. Technol.,* 2010, 54, 63–68, DOI:10.1007/s10971-010-2158-0.

55. Wu Y., Jia W., An Q., Liu Y., Chen J., and Li G., Multiaction antibacterial nanofibrous membranes fabricated by electrospinning: an excellent system for antibacterial applications, *Nanotechnology,* 2009, 20(24), 245101, DOI: 10.1088/0957-4484/20/24/245101.

56. Lee K. and Lee S., Multifunctionality of poly(vinyl alcohol) nanofiber webs containing titanium dioxide, *J. Appl. Polym. Sci.,* 2014, 124, 4038–4046.

57. Uykun N., Ergal I., Kurt H., Gokceoren A. T., Gocek I., Kayaoglu B. K., Akarsubasi A. T., and Sarac A. S., Fabrication of an antibacterial nanofibrous PVP/CTAB membrane for potential biomedical applications, *J. Bioact. Biocomp. Polym.,* 2014, 29, 4, 382–397.

58. Aussawasathien D., Dong J. H., and Dai L., Electrospun polymer nanofiber sensors, *Synth. Met.,* 2005, 154, 37–40.

59. Virji S., Huang J. X., Kaner R. B., and Weiller B. H., Polyaniline nanofiber gas sensors: examination of response mechanisms, *Nano Lett.,* 2004, 4, 491–496.

60. Lyons J. and Kaufmann J., Electrospinning: past, present & future, *Text. World,* 2004, 8, 46.

61. Solcova O., Balkan T., Guler Z., Morozova M. A., Dytrych P., and Sarac A. S., New preparation route of TiO_2 nanofibers by electrospinning: spectroscopic and thermal characterizations, *Sci. Adv. Mater.,* 2014, 6, 12, 2618–2624.

62. Liu S., Liu B., Nakata K., Ochiai T., Murakami T., and Fujishima A., Electrospinning preparation and photocatalytic activity of porous TiO_2 nanofibers, *J. Nanomat.,* 2012, Article ID 491927, 5 pages, DOI:10.1155/2012/491927.

63. Fong H. http://webpages.sdsmt.edu/~hfong/research.htm

64. Zhu H., Du M. L., Zhang M., Zou M. L., Yang T. T., Wang L. N. JuMing, Yao J. M., and Guo B. C., Probing the unexpected behavior of AuNPs migrating through nanofibers: a new strategy for the fabrication of carbon nanofiber–noble metal nanocrystal hybrid nanostructures, *J. Mater. Chem. A*, 2014, 2, 11728–11741, DOI: 10.1039/C4TA01624F.

65. Zussman E., Chen X., Ding W., Calabri L., Dikin D., Quintana J., and Ruoff R. S., Mechanical and structural characterization of electrospun PAN-derived carbon nanofibers, *Carbon*, 2005, 43, 2175.

66. Zhou Z., Lai C., Zhang L., Qian Y., Hou H., Reneker D. H., and Fong H., Development of carbon nanofibers from aligned electrospun polyacrylonitrile nanofiber bundles and characterization of their microstructural, electrical, and mechanical properties, *Polymer*, 2009, 50, 2999–3006.

67. Prilutsky S., Zussman E., and Cohen Y., The effect of embedded carbon nanotubes on the morphological evolution during the carbonization of poly(acrylonitrile) nanofibers, *Nanotechnology*, 2008, 19, 165603, DOI: 10.1088/0957-4484/19/16/165603.

68. Prilutsky S., Zussman E., and Cohen Y. J., Carbonization of electrospun poly(acrylonitrile) nanofibers containing multiwalled carbon nanotubes observed by transmission electron microscope with in situ heating, *Polym. Sci. B*, 2010, 48, 2121–2128.

69. Shim W. G., Kim C., Lee J. W., Yun J. J., Jeong Y. I., Moon H., and Yang K. S., Adsorption characteristics of benzene on electrospun derived porous CNFs, *J Appl Polym Sci.*, 2006, 102, 2454–2462.

70. Nataraj S. K., Yang K. S., and Aminabhavi T. M., Polyacrylonitrile-based nanofibers A state-of-the-art review, *Prog Polym. Sci.*, 2012, 37, 487–513.

71. Zussman E., Yarin A. L., Bazilevsky A. V., Avrahami R., and Feldman M., Electrospun polyacrylonitrile/poly(methyl methacrylate)-derived turbostratic carbon micro-/nanotubes, *Adv. Mater.*, 2006, 18, 348–353.

72. Kim C., Jeong Y. I., Ngoc B. T. N., Yang K. S., Kojima M., Kim Y. A., Endo M., and Lee J. W., Synthesis and characterization of porous carbon nanofibers with hollow cores through the thermal treatment of electrospun copolymeric nanofiber webs, *Small*, 2007, 3, 91–95.

73. Ju Y. W., Choi G. R., Jung H. R., and Lee W. J., Electrochemical properties of electrospun PAN/MWCNT carbon nanofibers electrodes coated with polypyrrole, *Electrochim Acta*, 2008, 53, 5796. http://dx.doi.org/10.1016/j. electacta. 2008.03.028.

74. Guo Q., Zhou X., Li X., Chen S., Seema A., Greiner A., and Hou H., Supercapacitors based on hybrid carbon nanofibers containing multi-walled carbon nanotubes, *J. Mater Chem.*, 2009, 19, 2810. http://dx.doi.org/10.1039/B820170F.

75. Lu X., Zhang W., Wang C., Wen T. C., and Wei Y., One-dimensional conducting polymer nanocomposites: synthesis, properties and applications, *Prog. Polym. Sci.*, 2011, 36, 5, 671–712, DOI:10.1016/j.progpolymsci.2010.07.010.

76. Guimard N. K., Gomez N., and Schmidt C. E., Conducting polymers in biomedical engineering, *Prog. Polym. Sci.*, 2007, 32, 8–9, 876–921, DOI:10.1016/j.progpolymsci.

77. Alemayehu T. and Sergawie A., The conducting biological pigment; melanin polymer, *IJTEEE*, 2014, 2(7), 2347–4289.

78. Osikowicz W., Denier van der Gon A. W., Crispin X., de Jong M. P., Friedlein R., Groenendaal L., Fahiman M., Beijonne D., Lazzaroni, R., and Salaneck W. R., A joint theoretical and experimental study on the electronic properties of phenyl-capped 3,4-ethylenedioxythiophene oligomers, *J. Chem. Phys.*, 2003, 119, 10415. http://dx.doi.org/10.1063/1.1617273.

79. Absorbtion and Emission. http://photonicswiki.org/index.php?title=Main_Page#Absorption_and_Emission_of_Light

80. Massonnet N, Carella A., Jaudouin O, Rannou P., Laval G., Celle C., and Simonato J. P., Improvement of the Seebeck coefficient of PEDOT:PSS by chemical reduction combined with a novel method for its transfer using free-standing thin films, *J. Mater. Chem. C*, 2014, 2, 1278–1283, DOI:10.1039/C3TC31674B.

81. Beverina L., Pagani G. A., and Sassi M., Multichromophoric electrochromic polymers: colour tuning of conjugated polymers through the side chain functionalization approach, *Chem. Commun.*, 2014, 50, 5413–5430, DOI:10.1039/C4CC00163J.

82. Jakobsson F. L. E., Crispin X., Lindell L., Kanciurzewska A., Fahlman M., Salaneck W. R., and Berggren M., Towards all-plastic flexible light emitting diodes, *Chem. Phys. Lett.*, 433, 2006, 110–114.

83. Shah A. A., Electrochemical synthesis and spectroelectrochemical characterization of conducting copolymers of aniline and *o*-aminophenol. http://www.qucosa.de/fileadmin/data/qucosa/documents/5387/data/anwar.pdf.

84. Apperloo J. J. and Janssen R. A. J., Solvent effects on the π-dimerization of cation radicals of conjugated oligomers, *Synth. Met.*, 100, 1999, 373.

85. Jiang X., Harima Y., Yamashita K., Tada Y., Ohshita J., and Kunai A., Doping-induced change of carrier mobilities in poly(3-hexylthiophene) films with different stacking structures, *Chem. Phys. Lett.*, 2002, 364, 5–6, 616–620.

86. Jang J., Conducting polymer nanomaterials and their applications, *Adv. Polym. Sci.*, 2006, 199, 189–259.

87. Waltman R. J. and Argon J. B., Electrically conducting polymers: a review of the electropolymerization reaction, of the effects of chemical structure on polymer film properties, and of applications towards technology, *Can. J. Chem.*, 1986, 64, 76–95.

88. Peng H., Zhang L., Soeller C., and Travas-Sejdi J., Review conducting polymers for electrochemical DNA sensing, *Biomaterials,* 2009, 30, 2132–2148.

89. Stenger-Smıth J. D., Intrinsically electrically conducting polymers. Synthesis, characterization, and their applications, *Prog. Polym. Sci.,* 1998, 23, 57–79.

90. Lange U., Roznyatovskaya N. V., and Mirsky V. M., Conducting polymers in chemical sensors and arrays, *Anal. Chim. Acta,* 2008, 6, 14, 1–26.

91. Gangopadhyay R. and De A., Conducting polymer nanocomposites: a brief overview, *Chem. Mater.,* 2000, 12, 608–622.

92. Wang Y. and Jing X., Intrinsically conducting polymers for electromagnetic interference shielding, *Polym. Adv. Technol.*, 2005, 16, 344–351.

93. Brar A. S. and Kaur J., 2D NMR studies of acrylonitrile–methyl acrylate copolymers. *Eur. Polym. J.*, 2005, 41, 10, 2278–2289, DOI:10.1016/j.eurpolymj.

94. Gurunathan K., Murugan A. V., Marimuthu R., Mulik, U., and Amalnerkar D., Electrochemically synthesised conducting polymeric materials for applications towards technology in electronics, optoelectronics and energy storage devices, *Mater. Chem. Phys.*, 1999, 61, 3, 173–191, DOI:10.1016/S0254-0584(99)00081-4.

95. Grozema F. C., Siebbeles L. D. A., Gelinck G, H., and Warman J. M., Molecular wires and electronics, *Top. Curr. Chem.*, 2005, 257, 135–164.

96. Laforgue A. and Robitaille L., Fabrication of poly-3-hexylthiophene/polyethylene oxide nanofibers using electrospinning, *Synth. Met.,* 2008, 158, 577–584.

97. Pinto N. J., Ramos I., Rojas R., Wang P.-C., Johnson A. T., Jr., Electric response of isolated electrospun polyaniline nanofibers to vapors of aliphatic alcohols, *Sensor Actuat B-Chem*, 129(2), 621–627.

98. Elsenbaumer, R. L., Jen, K.Y., Miller, G. G., Eckhardt, H., Shacklette, L. W., and Jow, R., 1987. Poly (alkyl thiophenes) and poly (substituted heteroaromatic vinylenes): versatile, highly conductive, processible polymers with tunable properties, in *Electronic Properties of Conjugated Polymers*, Kuzmany, H., Mehring, M., and Roth, S. (Eds.), Springer, Berlin. ISBN: 0-387-18582-8.
99. Polythiophene. https://en.wikipedia.org/wiki/Polythiophene
100. Dolas H. and Sarac A. S., An impedance-morphology study on poly(3-methylthiophene) coated electrode obtained in boron trifluoride diethyl etherate -acetonitrile, *Synthetic Metals* 2014,195, 44–53.
101. Corradi R. and Armes S. P., Chemical synthesis of poly(3,4-ethylenedioxythiophene), *Synt. Met.*, 1997, 84, 453–454.
102. Delzenne G. A. 1976. *Encyclopaedia of Polymer Science and Technology, Supplement*, Wiley, New York, Vol.1, p.401
103. Badhwar S., 2006. *Design, Fabrication and Characterisation of Organic Electrochemical Transistors*, Chapter 2, Material aspects of PEDOT:PSS.
104. TDA Research. Intrinsically conducting polymers. http://www.tda.com/eMatls/icp.htm
105. Peng C., Zhang S., Jewell D., and Chen G. Z., Carbon nanotube and conducting polymer composites for supercapacitors. *Prog. Nat. Sci.*, 2008, 18, 7, 777–788, DOI:10.1016/j.pnsc.2008.03.002.
106. Ramya R., Sivasubramanian R., and Sangaranarayanan M. V., Conducting polymers-based electrochemical supercapacitors: progress and prospects, *Electrochim. Acta*, 2013, 101, 109–12, DOI:10.1016/j.electacta.2012.09.116.
107. Elschner A., Kirchmeyer S., Lövenich W., Merker U., and Reuter K., 2011. *PEDOT Principles and Applications of an Intrinsically Conductive Polymer*, CRC Press, Boca Raton.
108. Bhat D. K., Kumar M. S., N and p doped poly(3,4-ethylenedioxythiophene) electrode materials for symmetric redox supercapacitors, *J. Mater. Sci.*, 2007, 42, 8158–8162.
109. Laforgue A., All-textile flexible super capacitors using electrospun poly(3,4-ethylenedioxythiophene) nanofibers, *J. Power Sources*, 2011, 196, 559–564.
110. Gaupp C. L., Welsh D. M., and Reynolds J. R., Poly(ProDOT-Et2): a high-contrast, high-coloration efficiency electrochromic polymer, *Macromol. Rapid Commun.*, 2002, 23, 885–889.

111. Beaujuge P. M. and Reynolds J. R., Color control in π-conjugated organic polymers for use in electrochromic devices., *Chem. Rev.*, 2010, 110, 268–320.

112. (a) *Handbook of Organic Conductive Materials and Polymers*, 1997. Nalwa H. S. (Ed.), Wiley, New York. (b) *Handbook of Conducting Polymers*, 2nd ed., 1998. Skotheim T. A., Elsenbaumer R. L., and Reynolds J. F. (Eds.), Marcel Dekker, New York. Coontz and B. Hanson, *Science*, 2004, 305, 957.

113. Groenendaal L., et al. PEDT and its derivatives: past, present, and future, *Adv. Mater.*, 2000, 12(7).

114. Havinga E. E., Mutsaers C. M., and Jenneskens L. W. Absorption properties of alkoxy-substituted thienylene–vinylene oligomers as a function of the doping level, *Chem. Mater.*, 1996, 8, 3, 769–776, DOI: 10.1021/cm9504551.

115. Schottland P., Stéphan O., Le Gall P. Y., and Chevrot C., Synthesis and polymerization of new monomers derived from 3, 4-ethylenedioxythiophene, *J. Chim. Phys.*, 1998, 95, 1258–1261.

116. Pettersson L. A. A., Johansson T., Carlsson F., Arwin H., Inganäs O., et al., Anisotropic optical properties of doped poly(3,4-ethylenedioxythiophene), *Synth. Met.*, 1999, 101, 198–199.

117. Bayer A. G., New polythiophene dispersions: their preparation and their use, *Eur. Patent*, 1991, 440957.

118. Wang C., Schindler J., Kannewurf C., and Kanatzidis. M., Poly(3,4-ethylenedithiathiophene). A new soluble conductive polythiophene derivative, *Chem. Mater.*, 1995, 7, 58–68.

119. Kudoh Y., Akami K., and Matsuya Y., Solid electrolytic capacitor with highly stable conducting polymer as a counter electrode, *Synth. Met.*, 1999, 102, 973–974.

120. Basak Demircioglu, MSc Thesis at Istanbul Technical University, Graduate School of Science, Eng & Tech. 2013.

121. Nanofibers and Nanocomposites Poly(3,4-ethylene dioxythiophene)/Poly(styrene sulfonate) by Electrospinning. PhD Thesis submitted to the faculty of Drexel University by Afaf Khamis El-Aufy, 2004.

122. Heeger A. J., Semiconducting and metallic polymers: the fourth generation of polymeric materials., *Angew. Chem., Int. Ed.*, 2001, 40, 2591–2611.

123. Hatchett D. W. and Josowicz M., Composites of intrinsically conducting polymers as sensing nanomaterials, *Chem. Rev.*, 2008, 108, 746–769.

124. Roncali J., Conjugated poly(thiophenes): synthesis, functionalization, and applications, *Chem. Rev.*, 1992, 92, 711–738.

125. Li C., Bai H., Shi G., Conducting polymer nanomaterials: electrosynthesis and applications. *Chem. Soc. Rev.,* 2009, 38, 2397–2409.

126. Li C. and Imae T., Electrochemical and optical properties of the poly(3,4-ethylenedioxythiophene) film electropolymerized in an aqueous sodium dodecyl sulfate and lithium tetrafluoroborate medium macromolecules, 2004, 37, 2411–2416.

127. Pringle J. M., Forsyth M., Wallace G. G., and MacFarlane D. R., Solution-surface electropolymerization: a route to morphologically novel poly(pyrrole) using an ionic liquid, *Macromolecules,* 2006, 39, 7193–7195.

128. Gao M., Huang S. M., Dai L. M., Wallace G., Gao R. P., and Wang Z. L., Aligned coaxial nanowires of carbon nanotubes sheathed with conducting polymers, *Angew. Chem., Int. Ed.*, 2000, 39, 3664–3667.

129. Hughes M., Shaffer M. S. P., Renouf A. C., Singh C., Chen G. Z., Fray J., and Windle A. H., Electrochemical capacitance of nanocomposite films formed by coating aligned arrays of carbon nanotubes with polypyrrole, *Adv. Mater.*, 2002, 14, 382–385.

130. Chen G. Z., Shaffer M. S. P., Coleby D., Dixon G., Zhou W. Z., Fray D. J., and Windle A. H., Carbon nanotube and polypyrrole composites: Coating and doping, *Adv. Mater.*, 2000, 12, 522–526.

131. Wu X. F. and Shi G. Q., Synthesis of a carboxyl-containing conducting oligomer and non-covalent sidewall functionalization of single-walled carbon nanotubes, *J. Mater. Chem.*, 2005, 15, 1833–1837.

132. Yerlikaya Y., Unsal C., and Sarac A. S., Nanofibers of poly(acrylonitrile-co-methylacrylate)/polypyrrole core–shell nanoparticles, *Adv. Sci. Eng. Medi.*, 2014, 6, 301–310.

133. Unsal C., Kalaoglu F., Karakas H., and Sarac A. S., Polypyrrole/poly(acrylonitrile-co-butyl acrylate), *Comp. Adv. Polym. Technol.*, 2013, 32, S1, E784–E792, DOI:10.1002/adv.21321.

134. Luo X. L., Killard A. J., and Smyth M. R., Nanocomposite and nanoporous polyaniline conducting polymers exhibit enhanced catalysis of nitrite reduction, *Chem. Eur. J.*, 2007, 13, 2138–2143.

135. Innis P. C., Norris I. D., Kane-Maguire L. A. P., and Wallace G. G., Electrochemical formation of chiral polyaniline colloids codoped with (+)- or (-)-10-camphorsulfonic acid and polystyrene sulfonate, *Macromolecules*, 1998, 31, 6521–6528.

136. Yoneyama H., Shoji Y., and Kawai K., Electrochemical synthesis of polypyrrole films containing metal oxide particles, *Chem. Lett.*, 1989, 1067–1070.

137. Drury A., Chaure S., Kröell M., Nicolosi, V. Chaure N., and Blau W. J., Fabrication and characterization of silver/polyaniline composite nanowires in porous anodic alumina, *Chem. Mater.*, 2007, 19, 4252–4258, DOI:10.1021/cm071102s.

138. Roux S., Soler-Illia G., Demoustier-Champagne S., Audebert P., and Sanchez C., Titania/polypyrrole hybrid nanocomposites built from in-situ generated organically functionalized nanoanatase building blocks, *Adv. Mater.*, 2003, 15, 217–221.

139. Yan H. L. and Shi G. Q., Incorporation of gold nanocrystals into poly(3-alkylthiophene) nanowires and fabrication of gold nanowires, *Nanotechnology*, 2006, 17, 13–18.

140. Tsakova V., How to affect number, size, and location of metal particles deposited in conducting polymer layers, *J. Solid State Electrochem.*, 2008, 12, 1421–1434.

141. Bartlett P. N. and Cooper J. M., A review of the immobilization of enzymes in electropolymerized films, *J. Electroanal. Chem.*, 1993, 362, 1–12.

142. Rahman M. A., Kumar P., Park D. S., and Shim Y. B., Electrochemical sensors based on organic conjugated polymers, *Sensors*, 2008, 8, 118–141.

143. Gao M., Dai L. M., and Wallace G. G., Biosensors based on aligned carbon nanotubes coated with inherently conducting polymers, *Electroanalysis*, 2003, 15, 1089–1094.

144. Liu H. S., Song C. J., Zhang L., Zhang J. J., Wang H. J., and Wilkinson D. P., A review of anode catalysis in the direct methanol fuel cell, *J. Power Sources*, 2006, 155, 95–110.

145. Li J. and Lin X. Q., A composite of polypyrrole nanowire platinum modified electrode for oxygen reduction and methanol oxidation reactions, *J. Electrochem. Soc.*, 2007, 154, B1074–B1079.

146. Huang X. H., Tu J. P., Xia X. H., Wang X. L., and Xiang J. Y., Nickel foam-supported porous NiO/polyaniline film as anode for lithium ion batteries, *Electrochem. Commun.*, 2008, 10, 1288–1290.

147. Zhao L., Tong L., Li C., Gu Z. Z., and Shi G. Q., Polypyrrole actuators with inverse opal structures, *J. Mater. Chem.*, 2009, 19, 1653–1658.

148. Wang J., Naguib H. E., and Bazylak A., Electrospun porous conductive polymer membrane., *Proc. SPIE* 8342 83420F-1; Behavior and mechanics of multifunctional materials and composites 2012, Goulbourne N. C., and Ounaies Z. (Eds.), *Proc. SPIE* 8342, 83420F · © 2012 SPIE, DOI: 10.1117/12.923599.

149. Hayati I., Eddies inside a liquid cone stressed by interfacial electrical shear, *Colloids and Surfaces*, 1992, 65, 77–84.

150. Reneker D. H., Yarin A. L., Fong H., Koombhongse S., Bending instability of electrically charged liquid jets of polymer solutions in electrospinning, *J. Appl. Phys.*, 2000, 87, 4531.

151. Koombhongse S., Liu W. X., Reneker, D. H., Flat polymer ribbons and other shapes by electrospinning, *J. Polym. Sci. Part B-Polym. Phys.* 2001, 39, 2598–2606.

152. Zeng J., Xu X. Y., Chen X. S., Liang Q. Z., Bian X.C., Yang L. X., and Jing, X. B., Biodegradable electrospun fibers for drug delivery, *J Control Release*, 2003, 92, 227–231.

153. Kim S. H., Nam Y. S., Lee T. S., and Park W. H., Silk fibroin nanofiber: electrospinning, properties, and structure, *Polym. J.*, 2003, 35, 185–190.

154. Zhang Y., Rutledge G. C., Electrical conductivity of electrospun polyaniline and polyaniline-blend fibers and mats, *Macromolecules*, 2012, 45, 10. 4238-4246, DOI: 10.1021/ma3005982.

155. Li C., Chartuprayoon N., Bosze W., Low K., Lee K. H., Nam J., and Myung N. V., Electrospun polyaniline/poly(ethylene oxide) composite nanofibers based gas sensor, *Electroanalysis*, 2014, 26, 711–722.

156. Liu S.-L., Long Y.-Z., Zhang Z.-H., Zhang H.-D., Sun B., Zhang J.-C., and Han W.-P., Assembly of oriented ultrafine polymer fibers by centrifugal electrospinning, *J. Nanomater.*, 2013, Article ID 713275, 9 pages http://dx.doi.org/10.1155/2013/713275.

157. Yousefzadeh M., Latifi M., Teo W.-E., Amani-Tehran M., and Ramakrishna S., Producing continuous twisted yarn from well-aligned nanofibers by a water vortex, Society of Plastics Engineers (SPE) *Plast. Res. Online*, 2011, DOI: 10.1002/spepro.003599.

158. Li X., Liu Y., Shi Z., Li C., and Chen G., Influence of draw ratio on the structure and properties of PEDOT-PSS/PAN composite conductive fibers, *RSC Adv.*, 2014,4, 40385–40389, DOI: 10.1039/C4RA05952B.

159. Jin L., Wang T., Feng Z. Q., Leach M. K., Wu J., Mo S., and Jiang Q., A facile approach for the fabrication of core–shell PEDOT nanofiber mats with superior mechanical properties and biocompatibility, *J. Mater. Chem. B*, 2013,1, 1818–1825, DOI: 10.1039/C3TB00448A.

160. Choi J., Lee J., Choi J., Jung D., and Shim S. E., Electrospun PEDOT:PSS/PVP nanofibers as the chemiresistor in chemical vapour sensing, *Synth. Met.*, 2010, 160, 1415–1421, DOI: 10.1016/j.synthmet.2010.04.021.

161. Rubinson J. F. and Kayinamura Y. P., Charge transport in conducting polymers: Insights from impedance spectroscopy, *Chem. Soc. Rev.*, 2009, 38, 3339–3347.

162. Glarum S. and Marshall J., Electron delocalization in poly(aniline), *J. Phys. Chem.*, 1988, 92, 4210–4217.

163. Albery W., Chen Z., Horrocks B., Mount A., Wilson P., Bloor D., Monkman A., and Elliott C., Spectroscopic and electrochemical studies of charge transfer in modified electrodes, Faraday discuss, *Chem. Soc.*, 1989, 88, 247–259.

164. Mathias M. and Haas O., An alternating current impedance model including migration and redox-site interactions at polymer-modifed electrodes, *J. Phys. Chem.*, 1992, 96, 3174–3182.

165. Tanguy J., Mermilliod N., and Hoclet M., Capacitive charge and noncapacitive charge in conducting polymer electrodes, *J. Electrochem. Soc.*, 1987, 134, 795–802.

166. Lang G. and Inzelt G., Some problems connected with impedance analysis of polymer film electrodes: effect of the film thickness and the thickness distribution, *Electrochim. Acta*, 1991, 36, 847–854.

167. Rammelt U. and Reinhard G., On the applicability of a constant phase element of solid metal electrodes, *Electrochim. Acta*, 1990, 35, 1045–1049.

168. Ho C., Raistrick I., and Huggins R., Application of AC techniques to the study of lithium diffusionin tungsten trioxide thin films, *J. Electrochem. Soc.*, 1980, 127, 343–350.

169. Wan, M., Polyaniline as a promising conducting polymer, In: *Conducting Polymers with Micro or Nanometer Structure,* Springer e book, ISBN 978-3-540-69322-2. Springer Berlin Heidelberg. 2008

170. Leclerc M., Guay J., and Dao L. H., Synthesis and characterization of poly(alkylanilines), *Macromolecules*, 1982, 22, 649–653.

171. Wei Y., Focke W. W., Wnek G. E., Ray A., and MacDiarmid A. G., Specialty polymers: materials and applications, *J. Phys. Chem.*, 1989, 93, 495–499.

172. MaCinnes D. Jr., and Funt B. L., Poly-o-methoxyaniline: a new soluble conducting polymer, *Synth. Met.*, 1988, 25, 235–242.

173. Mattoso L. H. C., Mello S. V., Riul A. Jr., Oliverira O. N. Jr., and Faria R. M., Synthesis and characterization of poly(o-phenetidine) for the fabrication of Langmuir and Langmuir-Blodgett films, *Thin Solid Films*, 1994, 244, 714–717.

174. Koenhen D. M. and Smolders C. A., The determination of solubility parameters of solvents and polymers by means of correlations with other physical quantities, *J. Appl. Polym. Sci.*, 1975, 19, 1163–1179.

175. Patent: Solvent Containing Carbon Nanotube Aqueous Dispersions EP 1910224 A1 (WO2008002317A1).

176. Gupta B. and Prakash R., Synthesis of functionalized conducting polymer "polyanthranilic acid" using various oxidizing agents and formation of composites with PVC, *Polym. Adv. Technol.*, 2011, 22, 1982–1988.

177. Yue J. and Esptein A. J., Conducting polymers with micro or nanometer structure, *J. Am. Chem. Soc.*, 1990, 112, 2800.

178. Bhattacharya A. and de Amitabha, Conducting polymers in solution: progress toward processibility, J.M.S., *Rev. Macromol. Chem. Phys.*, 1999, C39, 1, 17–56.

179. Nguyen M. T. and Diaz A. F., Synthesis of new water-soluble self-doped polyaniline, *Macromolecules*, 1995, 28, 3411.

180. Yan H., Wang H. J., Adisasmito S., and Toshima N., Novel syntheses of poly(*o*-aminobenzoic acid) and copolymers of *o*-aminobenzoic acid and aniline as potential candidates for precursor of polyaniline, *Bull. Chem. Soc. Jpn.*, 1996, 69, 2395.

181. Dash M. P., Tripathy M., Sasmal A., Gourang C., Mohanty C., and Nayak P. L., Poly anthranilic acid/multi-walled carbon nanotube composites: spectral, morphological, and electrical properties, *J. Mater. Sci.*, 2010, 45, 3858–3865.

182. Yue J. and Epstein A. J., Conducting polymers with micro or nanometer structure, *J. Chem. Soc. Chem. Commun.*, 1992, 21, 1540–1542.

183. Yue J., Wang Z. H., Cromack K. R., Epstein A. J., and MacDiarmid A. G., Organic electronic materials: conjugated polymers and low molecular weight, *J. Am. Chem. Soc.*, 1991, 113, 2665–2671.

184. Wang X. H., Li J., Wang L. X., Jing X. B., and Wang F. S., Synthesis and properties of poly(aniline-co-anisidine), *Synth. Met.*, 1995, 69, 145–146.

185. Yamamoto K. and Taneichi D., Electrochemical catalytic reduction of oxygen by a self-doped polyaniline-co-porphyrin complex-modified glassy carbon electrade, *Macromol. Chem. Phys.*, 2000, 201, 6–11.

186. Kilmartin P. A. and Wright G. A., Conducting polymers as free radical scavengers, *Synth. Met.*, 1997, 88, 153–162.

187. Li C. and Mu S., Conducting polymers of aniline I. Electrochemical synthesis of a conducting composite, *Synth. Met.*, 2004, 144, 143–149.

188. Karyakin A. A., Maltsev I. A., and Lukachova L. V., The influence of defects in polyaniline structure on its electroactivity: Optimization of 'self-doped' polyaniline synthesis, *J. Electroanal. Chem.*, 1996, 402, 217–219.

189. Nateghi M. R. and Borhani M., Preparation, characterization and application of poly anthranilic acid-co-pyrrole, *React. Funct. Polym.*, 2008, 68, 153–160.

190. Giray D., Balkan T., Dietzel B., and Sarac A. S., Electrochemical impedance study on nanofibers of poly(m-anthranilic acid)/polyacrylonitrile blends, *Eur. Polym. J.*, 2013, 49, 2645–2653, doi.org/10.1016/j.eurpolymj.2013.06.012.

191. Jacob M. M. E., and Arof A. K., FTIR studies of DMF plasticized polyvinyledene fluoride based polymer electrolytes, *Electrochim. Acta*, 2000, 45, 17021–1706.

192. Li C., Bai H., and Shi G. Q, Conducting polymer nanomaterials: Electrosynthesis and applications, *Chem. Soc. Revi.*, 2009, 38, 2397–2409.

193. Gupta V. and Miura N., High performance electrochemical supercapacitor from electrochemically synthesized nanostructured polyaniline, *Mater. Lett.*, 2006, 60, 1466–1469.

194. Laith Al-Mashat and Debiemme-Chouvy C., Electropolymerized polypyrrole nanowires for hydrogen gas sensing, *J. Phys. Chem. C*, 2012, 116(24), 13388.

195. Kim J. H., Sharma A. K., Lee Y. S., Synthesis of polypyrrole and carbon nano-fiber composite for the electrode of electrochemical capacitors, *Materi. Lett.*, 2006, 60, 1697–1701.

196. Ashok K., Sharma S. K., Kim J. H, and Lee Y. S., An efficient synthesis of polypyrrole/carbon fiber composite nano-thin films, *Int. J. Electrochem. Sci.*, 2009, 4, 1560–1567.

197. Islam M. M., Chidembo A. T., Aboutalebi S. H., Cardillo D., Liu H. K., Konstantinov K., and Dou S. X., Liquid crystalline graphene oxide/PEDOT-PSS self-assembled 3D architecture for binder-free supercapacitor electrodes, *Frontiers Energy Res.*, 2014, 2, Article 31, 1–10.

198. Noh K. A., Kim D. W., Jin C. S. B., Shin K. H., Kimb J. H., and Ko J. M., Synthesis and pseudo-capacitance of chemically-prepared polypyrrole powder, *J. Power Sources*, 2003, 124, 593–595.

199. Khomenko V., Frackowiak E., and B´eguin F., Determination of the specific capacitance of conducting polymer/nanotubes composite

electrodes using different cell configurations, *Electrochim. Acta*, 2005, 50, 2499–2506.

200. Park J. H., Ko J. M., Parka O. O., and Kim D. W., Capacitance properties of graphite/polypyrrole composite electrode prepared by chemical polymerization of pyrrole on graphite fiber, *J. Power Sources*, 2002, 105, 20–25.

201. Raicopol M., Pruna A., and Pilan L., Supercapacitance of single-walled carbon nanotubes-polypyrrole composites, *J. Chem.*, 2013, Article ID 367473, 7 pages, 2013, DOI:10.1155/2013/367473.

202. Musiani M. M., Characterization of electroactive polymer layers by electrochemical impedance spectroscopy (EIS), *Electrochim. Acta*, 1990, 35, 1665–1670.

203. Bobacka J., Lewenstam A., and Ivaska A., Electrochemical impedance spectroscopy of oxidized poly(3,4-ethylenedioxythiophene) film electrodes in aqueous solutions, *J. Electroanal. Chem.*, 2000, 489, 17–27.

204. Yasri N., Sundramoorthy A. K., Chang W.-J., and Gunasekaran S., Highly selective mercury detection at partially oxidized graphene/poly(3,4-ethylenedioxythiophene):poly(styrenesulfonate) nanocomposite film-modified electrode, *Front. Mater.*, 2014, DOI: 10.3389/fmats.2014.00033.

205. Gallegos A.K.C. and Rincon M., Carbon nanofiber and PEDOT-PSS bilayer systems as electrodes for symmetric and asymmetric electrochemical capacitor cells, *J. Power Sources*, 2006, 162, 743–747.

206. Long Y. Z., Li M. M., Gu C., Wan M., Duvail J. L., Liu Z., and Fan Z., Recent advances in synthesis, physical properties and applications of conducting polymer nanotubes and nanofibers, *Prog. Polym. Sci.*, 2011, 36, 1415–1442.

207. Hong K. H. and Kang T. J., Polyaniline–nylon 6, composite nanowires prepared by emulsion polymerization and electrospinning process, *J. Appl. Polym. Sci.*, 2006, 99, 1277–1286.

208. Sujith K., Asha A. M., Anjali P., Sivakumar N., Subramanian K. R. V., Nair S. V., and Balakrishnan A., Fabrication of highly porous conducting PANI-C composite fiber mats via electrospinning, *Mater. Lett.*, 2012, 67, 376–378.

209. Granato F., Bianco A., Bertarelli C., and Zerbi G., Composite polyamide 6/polypyrrole conductive nanofibers, *Macromol. Rapid Commun.*, 2009, 30, 453–458.

210. Rahy A. and Yang D. J., Synthesis of highly conductive polyaniline nanofibers, *Mater. Lett.,* 2008, 62, 4311–4314.

211. Silva A. B. and Bretas R. E. S., Preparation and characterization of PA6/PAni-TSA nanofibers, *Synth. Met.,* 2012, 162, 1537–1545.

212. Yu Q. Z., Shi M. M., Deng M., Wang M., and Chen H. Z., Morphology and conductivity of polyaniline sub-micron fibers prepared by electrospinning, *Mater. Sci. Eng. B,* 2008, 70–76.

213. Babel A, Li D., Xia Y., and Jenekhe S.A., Electrospun nanofibers of blends of conjugated polymers: Morphology, optical properties, and field-effect transistors, *Macromolecules*, 2005, 38, 4705–4711.

214. Cardenas J. R., Franca M. G. O., Vasconcelos E. A., Azevedo W. M., and Silva E. F., Growth of sub-micron fibres of pure polyaniline using the electrospinning technique, *J. Phys. D: Appl. Phys.*, 2007, 40, 1068–1071.

215. Choi S. S., Chu B. Y., Hwang D. S., Lee S. G., Park W. H., and Park J. K., Preparation and characterization of polyaniline nanofiber webs by template reaction with electrospun silica nanofibers, *Thin Solid Films,* 2005, 477, 233–239.

216. Neuberta S., Pliszkaa D., Thavasia V., Wintermantel E., and Ramakrishnaa S., Conductive electrospun PANi-PEO/TiO2 fibrous membrane for photo catalysis, *Mater. Sci. Eng. B,* 2011, 176, 640–646.

217. Li M., Guo Y., Wei Y., MacDiarmid E. G., and Lelkes P. I., Electrospinning polyaniline-contained gelatin nanofibers for tissue engineering applications, *Biomaterials*, 2006, 27, 2705–2715.

218. Fryczkowskia R. and Kowalczyk T., Nanofibres from polyaniline/polyhydroxybutyrate blends, *Synth. Met.,* 2009, 159, 2266–2268.

219. Yu X., Li Y., Zhu N., Yang Q., and Kalantarzadeh K., A polyaniline nanofibre electrode and its application in a self-powered photoelectrochromic cell, *Nanotechnology,* 2007, 18, 015201.

220. Hong K. H., Oh K. W., and Kang T. J., Preparation of conducting nylon-6 electrospun fiberwebs by the in situ polymerization of polyaniline, *J. Appl. Polym. Sci.,* 2005, 96, 983–991.

221. Lin J. Y., Cai Y., Wang X. F., Ding B., Yu J. Y., and Wang M. R., Fabrication of biomimetic superhydrophobic surfaces inspired by lotus leaf and silver ragwort leaf, *Nanoscale,* 2011, 3, 1258–1262, DOI:10.1039/c0nr00812e.

222. Asmatulu R., Ceylan M., and Nuraje, N., Study of superhydrophobic electrospun nanocomposite fibers for energy systems, *Langmuir,* 2011, 27, 504–507, DOI:10.1021/la103661c.

223. Wen S. P., Liu L., Zhang L. F., Chen Q., Zhang L. Q., and Fong H., Hierarchical electrospun SiO_2 nanofibers containing SiO_2 nanoparticles with controllable surface-roughness and/or porosity, *Mater. Lett.,* 2010, 64, 1517–1520, DOI:10.1016/j.matlet.2010.04.008.

224. Chen X., Dong B., Wang B. B., Shah R., and Li C.Y., Crystalline block copolymer decorated, hierarchically ordered polymer nanofibers, *Macromolecules,* 2010, 43, 9918–9927, DOI:10.1021/ma101900n.

225. Lai C., Guo Q. H., Wu X. F., Reneker D. H., and Hou H., Growth of carbon nanostructures on carbonized electrospun nanofibers with palladium nanoparticles, *Nanotechnology,* 2008, 19, 195303–195309, DOI:10.1088/0957-4484/19/19/195303.

226. Hou H. Q. and Reneker D. H., Carbon nanotubes on carbon nanofibers: a novel structure based on electrospun polymer nanofibers, *Adv. Mater.,* 2004, 16, 69–73, DOI:10.1002/adma.200306205.

227. Zander, N. E., Hierarchically structured electrospun fibers, *Polymers,* 2013, 5, 19–44.

228. Chronakis I. S., Grapenson S., and Jakob A., Conductive polypyrrole nanofibers via electrospinning: electrical and morphological properties, *Polymer,* 2006, 47, 1597–1603.

229. Arman B. and Sarac A. (2014) In situ preparation of core shell-polypyrrole/poly (acrylonitrile-co-vinyl acetate) nanoparticles and their nanofibers, *Soft Nanosci. Lett.* 4, 42–49, DOI: 10.4236/snl.2014.42006.

230. Chen R., Zhao S., Han G., and Dong J., Fabrication of the silver/polypyrrole/polyacrylonitrile composite nanofibrous mats, *Mater. Lett.,* 2008, 62, 4031–4034.

231. Cetiner S., Kalaoglu F., Karakas H., and Sarac A. S., Electrospun nanofibers of polypyrrole-poly(acrylonitrile-co-vinyl acetate), *Text. Res. J.,* 2010, 80, 17, 1784–1792.

232. Cetiner S., Sen S., Arman B., and Sarac A. S., Acrylonitrile/vinyl acetate copolymer nanofibers with different vinylacetate content, *J. Appl. Polym. Sci.,* 2013, 127, 5, 3830–3838, DOI:10.1002/app.37690.

233. Kai D., Prabhakaran M. P., Jin G., and Ramakrishna S., Polypyrrole-contained electrospun conductive nanofibrous membranes for cardiac tissue engineering, *J. Biomed. Mater. Res. A,* 2011, 99, 3.

234. Yanilmaz M., Kalaoglu F., Karakas H., and Sarac A. S., Preparation and characterization of electrospun polyurethane–polypyrrole nanofibers and films, *J. Appl. Polym. Sci.,* 2012, 125, 4100–4108.

235. UIB and Bursa Textile & Confection Research Center, The Fiber Society Spring 2010 International Conference, May 12—14, 2010, http://www.thefibersociety.org/Portals/0/Past%20Conferences/2010_Spring_Abstracts.pdf

236. Wen T. C., Luo S. S., and Yang C. H., Ionic conductivity of polymer electrolytes derived from various diisocyanate-based waterborne polyurethanes, *Polymer*, 2000, 41, 6755–6764.

237. Picciani P. H. S., Medeiros E. S., Pan Z., Wood D. F., Orts W. J., Mattoso L. H. C., and Soares B. G., Structural, electrical, mechanical, and thermal properties of electrospun poly(lactic acid)/polyaniline blend fibers, *Macromol. Mater. Eng.*, 2010, 295, 618–627.

238. Picciani P. H. S., Medeiros E. S., Pan Z., Orts W. J., Mattoso L. H. C., and Soares B. G., Development of conducting polyaniline/poly(lactic acid) nanofibers by electrospinning, *J. Appl. Polym. Sci.*, 2009, 112, 744–753.

239. Norris I. D., Shaker M. M., Ko F. K., and MacDiarmid A. G., Electrostatic fabrication of ultrafine conducting fibers: polyaniline polyethylene oxide blends, *Synth. Met.*, 2000, 114, 109–114.

240. Camposeo, A. and Pisignano, D., New perspectives for light-emitting polymer nanofibers. http://spie.org/newsroom/technical-articles-archive/3450-new-perspectives-for-light-emitting-polymer-nanofibers?highlight=x2402&ArticleID=x44148

241. Camposeo A. and Pisignano D., New perspectives for light-emitting polymer nanofibers. http://spie.org/newsroom/technical-articles-archive/3450-new-perspectives-for-light-emitting-polymer-nanofibers.

242. Fasano, V., Polini, A., Morello, G., Moffa, M., Camposeo, A., and Pisignano, D., Bright light emission and waveguiding in conjugated polymer nanofibers electrospun from organic salt added solutions, *Macromolecules*, 2013, 46, 5935–5942.

243. Pagliara S., Camposeo A., Mele E., Persano L., Cingolani R., and Pisignano D., Enhancement of light polarization from electrospun polymer fibers by room temperature nanoimprint lithography, *Nanotechnology*, 2010, 21, 215304, DOI:10.1088/0957-4484/21/21/215304.

244. Di Benedetto F., Camposeo A., Pagliara S., Mele E., Persano L., Stabile R., Cingolani R., and Pisignano D., Patterning of light emitting conjugated polymer nanofibers, *Nat. Nanotech.*, 2008, 3, 614–619, DOI:10.1038/nnano.2008.232.

245. Lee J. Y., Bashur C. A., Goldstein A. S., and Schmidt C. E., Polypyrrole-coated electrospun PLGA nanofibers for neural tissue applications, *Biomaterials,* 2009, 30, 4325–4335.

246. Breukers R. D., Gilmore K. J., Kita M., Wagner K. K., Higgins M. J., Moulton S. E., Clark G. M., Officer D. L., Kapsa R. M. I., and Wallace G. G., Creating conductive structures for cell growth: growth and alignment of myogenic cell types on polythiophenes, *J. Biomed. Mater. Res. A,* 2010, 95A, 1,256–268.

247. Zhang Y., Lim C. T., Ramakrishna S., and Minghuang Z., Recent development of polymer nanofibers for biomedical and biotechnological applications, *J. Mater. Sci.: Mater. Med.,* 2005, 16, 933–946.

248. Bendrea A. D., Cianga L., and Cianga I., Review paper: progress in the field of conducting polymers for tissue engineering applications, *J. Biomater. Appl.* 2011, 26.

249. Choi J., Lee J., Choi J., Jung D., and Shim S. E., Electrospun PEDOT:PSS/PVP nanofibers as the chemiresistor in chemical vapour sensing, *Synth. Met.,* 2010, 160, 1415–1421.

250. Macagnano A., Zampetti E., Pantalei S., Cesare F. D., Bearzotti A., and Persaud K. C., Nanofibrous PANI-based conductive polymers for trace gas analysis, *Thin Solid Films,* 2011, 520, 978–985.

251. Lin Q., Li Y., and Yang M., Polyaniline nanofiber humidity sensor prepared by electrospinning, *Sens. Actuators,* 2012, 161, 967– 972.

252. Miao Y. E., Fan W., Chen D., and Liu T., High-performance supercapacitors based on hollow polyaniline nanofibers by electrospinning, *ACS Appl. Mater. Interfaces,* 2013, 5, 10, 4423–4428, DOI: 10.1021/am4008352.

Index

AAO, *see* anodic alumunium oxide
AB, *see* Alamar Blue
acetone 14, 17, 19, 48, 52, 66, 98, 115, 116, 158, 170, 178, 180, 219, 250
acetonitrile 19, 96, 98, 99, 121, 122, 179, 206
AFM, *see* atomic force microscopy
Alamar Blue (AB) 221
alkyl-substituted polythiophenes 93, 95, 97, 99
aniline 4, 92, 111, 120–122, 124, 179, 183, 216, 217, 220, 241, 242, 251
aniline-doped silica nanofibers 220, 242
anodic alumunium oxide (AAO) 125
antimicrobials 54
atomic force microscopy (AFM) 95
axisymmetric instabilities 37

BAYTRON P 113
BDS, *see* broadband dielectric/impedance spectroscopy
benzene 18, 68, 121, 179, 181, 184, 187
Berry number 27, 30, 31, 33, 34, 134
biopolymers 5
biosensors 86, 91, 101, 210, 250
bipolarons 72, 74, 82, 83
blend fibers 92, 138, 219, 221
  electrospun PANI 138
BMSCs, *see* bone marrow stromal cells
Bode magnitude 174, 191, 194
broadband dielectric/impedance spectroscopy (BDS) 168

capacitors 61, 62, 101, 106, 107, 113–115, 117, 125, 162, 164, 197, 198, 210, 213, 242, 251
  double-layer 106, 107, 197
  solid electrolytic 114
carbon 3, 46, 47, 61, 63, 65–69, 95–99, 122, 126, 135, 136, 197, 203, 208, 209, 212, 213, 218, 223–225
carbon fibers 3, 46, 65, 66, 68, 69, 203
  nanoscaled 65, 66
  vapor-grown 203
carbonization 65, 66, 68, 69
carbon nanofiber precursor 40
carbon nanofibers 3, 66–69, 136, 197, 212, 213, 223, 224
  hollow 68, 213
carbon nanofibers (CNFs) 66
carbon nanospheres 63
carbon nanostructures 223, 224
carbon nanotubes (CNTs) 122, 218
carbon tetrachloride 97, 179
carboxylic acid 38, 182, 183, 187
carrier polymer 2, 216, 225
  insulating 216
cerium 116, 168, 170, 172, 229, 242

charge carriers 25, 72–74, 83, 84, 89, 119, 159, 164, 218, 224
charge mobilities 89
charge transfer 72, 74, 161, 162, 164–167, 185, 186, 191, 195, 210–212
chemical doping 78, 85
chitosan 40, 41
chloroform 14, 18, 90, 95, 97, 112, 134, 179, 217, 220, 224
CNFs, *see* carbon nanofibers
CNTs, *see* carbon nanotubes
composite fibers 92, 136, 146, 170, 195, 218, 220, 225
  conductive 92, 136, 146, 218, 220
composite materials 2, 10, 135, 206
composite nanofiber electrode 251
composite nanofibers 139, 146, 149, 153, 158, 171, 174, 194, 195, 218, 225, 228–231, 233, 235–237, 239, 242
  conductive 147, 226, 228, 230
conducting polymers (CPs) 71, 149
conduction bands 164
conductive nanofibers 215, 216, 218, 220, 222, 224, 226–228, 230, 232, 234, 236, 238, 240, 242, 244, 252
conductive nanofibrils 42, 43
conductive polymers 3, 72, 86, 89, 128, 153, 225
conjugated polymers 71–74, 76–80, 82–92, 94–96, 98, 100, 102, 111, 118, 119, 161, 164, 197–202, 210, 215, 244
  doped 71, 79, 82, 83, 88, 92, 111, 164, 197, 210
  neutral 71, 73, 74, 78–80, 83, 100, 119
  reduced neutral 79
conjugation 72, 77, 89, 176, 185, 186, 219
constant phase element (CPE) 164
CPE, *see* constant phase element
CPMEs, *see* conjugated polymer modified electrode
CPs, *see* conducting polymers
CV, *see* cyclic voltammogram
cyclic voltammetry 190, 205, 206
cyclic voltammogram (CV) 201

dielectric constant 15–17, 20, 21, 27, 42, 159, 238, 242
diethyl ether 14, 97, 99
differential scanning calorimetry (DSC) 39, 230
differentiated primary myotubes 248, 249
direct methanol fuel cells (DMFCs) 3, 101
DMA, *see* dynamic mechanical analysis
DMFCs, *see* direct methanol fuel cells
DNA 5, 123, 124, 135, 136
drug delivery system 48
DSC, *see* differential scanning calorimetry
dynamic mechanical analysis (DMA) 230

ECM, *see* equivalent circuit model
ECs, *see* electrochemical capacitors
EDLC, *see* electric double-layer capacitor

EIS, *see* electrochemical impedance spectroscopy
electric double-layer capacitor (EDLC) 69
electroactive fibers 44, 91, 217, 219
  composite 217
electroactive polymers 44, 89, 175
electrochemical activity 101, 182, 183, 190
electrochemical capacitors (ECs) 213
electrochemical codeposition 123, 124
electrochemical deposition 125, 198, 202
electrochemical impedance spectroscopy (EIS) 61, 107, 210
electrochemical polymerization 85, 96, 98–100, 108, 112, 119–121, 124, 127, 208
electron paramagnetic resonance (EPR) 83
electro-oxidation 101, 126
electropolymerization 99, 119, 122, 124, 126, 127, 197, 202
electrospinning 1, 2, 4–13, 15–17, 19–27, 33, 34, 40–44, 58, 59, 89–92, 116–118, 131–146, 152, 184, 185, 217–219, 227–231, 250–252
  coaxial 10, 42, 43, 90, 128, 139
electrospinning jet 16, 17, 24, 27, 42
electrospinning solutions 56, 116, 170, 225, 230
electrospun carbon/graphite nanofibers 65, 67, 69
electrospun carbon nanofibers 69, 223
electrospun ceramic nanofibers 63
electrospun fibers 5, 7, 20, 22, 31, 33, 40, 136, 138, 144, 222, 225, 228, 246, 248, 249
  aligned 144, 249
  nanosized 44
electrospun nanofiber equipment 53
electrospun nanofiber mats 56
electrospun nanofibers 2, 10, 38, 40, 42, 54, 61, 151, 168, 170, 171, 173, 174, 189–191, 218–221, 245, 246, 250, 251
  blended 219
  composite 2, 10, 136, 150, 168, 170, 171, 174, 184, 221, 239, 250, 251
  electroactive 89, 191, 219
  light-emitting 245, 246
electrospun PANI 91, 138, 139, 141
electrospun PANI–PMMA core–shell fibers 139
electrospun PEDOT-PSS/PVP nanofibers 157
electrospun polyaniline 3, 140
electrospun polymer fibers 17, 248
electrospun polymer nanofibers 53, 65, 135, 136
electrospun polymers 47, 59
electrosynthesis 123, 124, 126, 198
EPR, *see* electron paramagnetic resonance

equivalent circuit model (ECM) 161
ethanol 12, 14, 18, 52, 63, 92, 101, 158, 178, 180, 250
expanded-coil conformations 102

fibers 1–3, 5–10, 16, 17, 20–25, 44, 65–70, 90–93, 129, 130, 135–140, 143, 144, 146, 147, 217–222, 225, 239–241, 246–249
  glass 46, 129, 135, 217, 220, 221
  graphite 65
  hollow polypyrrole 225
  Kevlar 135
  light-emitting 246, 247
  pure polyaniline 219
  pure PVAc 151, 152, 171
Fourier transform infrared spectroscopy–attenuated total reflectance (FTIR-ATR) 38
FTIR-ATR, *see* Fourier transform infrared spectroscopy–attenuated total reflectance
fuel cells 3, 11, 101, 251

gas sensors 126, 250
GCEs, *see* glassy carbon electrodes
gelatin 92, 220, 221, 229
gelatin fibers 220, 221
gelation 3, 4, 64
gel permeation chromatography (GPC) 112
gel polymer electrolyte 99, 127
glassy carbon electrodes (GCEs) 212
glucose 126, 181
GPC, *see* gel permeation chromatography

HEPA, *see* high-efficiency particulate air
highest occupied molecular orbital (HOMO) 72
HOMO, *see* highest occupied molecular orbital
hydrogen-bonded carbonyls 230, 232
hydrogen bonding 20, 190, 232–234
hydroxypropylcellulose 15

indium tin oxide (ITO) 117, 122
ITO, *see* indium tin oxide

leucoemeraldine 175, 177, 183, 216
Lewis acid 99
lowest unoccupied molecular orbital (LUMO) 72
LUMO, *see* lowest unoccupied molecular orbital

metanilic acid 181, 182
methylene chloride 36, 179
MEV, *see* minimum electrospinning voltage
micro emulsion polymerization 124
minimum electrospinning voltage (MEV) 33

nanocomposites 4, 46, 62, 88, 119, 123–126, 135, 148, 153, 171, 197, 198, 212, 250
  inorganic/PANI 4
nanofiber mats 1, 17, 40, 56, 93, 146, 151, 171, 173, 224, 239, 241, 244, 252, 253
  composite 146, 171, 172, 239, 241

electrospun nylon-6 56
nanofiber meshes 10
nanofibers 1–7, 9–12, 36–38, 40–43, 45–71, 91–93, 135–139, 151–153, 156–159, 161, 162, 170–176, 184–197, 212, 213, 215–248, 250–252
as-spun 63, 67, 70, 138, 156, 189
bicomponent 10
blend 49, 50, 92, 138, 139, 191, 195, 217, 219–221, 229, 242
carbon/graphite 65, 67, 69
ceramic 63, 65
chitosan 40, 41
core–shell 42, 43, 68, 123
electrospun metallic 65
electrospun PVA 56, 57
electrospun SiO2 65, 222
gelatin 92, 220, 221, 229
light-emitting 71, 217, 244–247
piezoelectric 11
polyacrylonitrile 5, 66, 136, 184, 227, 228
porous graphite 65
porous silver 65
silica 51, 219, 220, 223, 242
thiol-functionalized 65
nanofiber webs 10, 54–56, 219, 242
electrospun TiO2/PVA 56
hybrid 219, 242
NMR, *see* nuclear magnetic resonance
nuclear magnetic resonance (NMR) 38, 94, 138
Nyquist plots 163, 195, 205

oligomers 75, 83, 99, 120, 225, 241
oxidative polymerization 4, 96, 97, 99, 111, 216, 229, 242
chemical 4, 96, 111, 216

PAN, *see* polyacrylonitrile
PANA, *see* poly(anthranilic acid)
PANA nanofibers 187, 188
PANA/PAN blends 184, 189–191, 195
PANA/PAN electrospun nanofibers 184
PANA/PAN fibers 189
PANA/PAN nanofibers 186, 189, 190
PAN-based carbon nanofiber bundles 66
PAN-based carbon nanofibers 67, 69
PAN-based nanofiber web 68
PANFs, *see* polyacrylonitrile nanofibers
PANI, *see* polyaniline
PANI fibers 93, 139, 217, 240
PANI films 191
PANI–gelatin blend fibers 92
PANI nanofibers 139, 220, 250, 251
hollow 251
PANI nanoparticles 220, 241
PANI nanowires 198, 199, 201
PAN nanofibers 55, 56, 65, 66, 185, 186, 189, 190
PEDOT 74, 75, 77–83, 85, 87, 91–93, 100–105, 107, 112–117, 146–159, 168–175, 203, 205, 210–213, 247, 248, 250
neutral polymer 113
PEDOT derivatives 101, 107, 112, 113
PEDOT films 101, 113, 210
PEDOT nanofiber mats 146

PEDOT nanotubes 92, 93
PEMFC, *see* proton exchange membrane fuel cell
PEO, *see* poly(ethylene oxide)
PEO blend fibers 138
PEO fibers 239
PEO nanofibers 12, 141, 228
  smooth 12
pernigraniline 175, 177, 183, 190, 216
poly(anthranilic acid) (PANA) 182
poly(ethylene oxide) (PEO) 5, 217
poly(vinyl acetate) (PVAc) 2, 101, 114, 115, 146
poly(vinyl alcohol) (PVA) 5, 92
poly(vinyl butyral) (PVB) 250
poly(vinyl chloride) (PVC) 40, 222
poly(vinylidene fluoride) (PVDF) 90, 136, 218
polyacetylene 73, 74, 87, 88, 95, 118, 213
  doped 88
polyacrylates 103
polyacrylonitrile (PAN) 5, 54, 91, 136, 146, 184
polyacrylonitrile nanofibers (PANFs) 66
polyalkylthiophene 93
polyaniline (PANI) 3, 4, 85, 87, 111, 118, 139, 198, 215
polybenzimidazole 136
polydimethylsiloxane 103
polyester 5, 232
polyethers 103
polyethylene 53, 81, 117, 132, 133, 220, 228
polyethylene oxide 133, 220, 228
polymer batteries 223
polymer-coated electrodes 108
polymer electrolyte 99, 127, 128, 210
polymer fibers 1, 17, 44, 144, 218, 244, 248
  conjugated 218, 244
polymeric materials 58, 71
  conjugated 71
polymeric nanofibers 4, 61, 161, 162, 164, 166, 168, 170, 172, 174, 176, 178, 180, 182, 184, 186
polymerization 96–101, 107, 108, 111–116, 118–122, 124, 126–128, 130, 203, 206, 208, 209, 215, 216, 220, 227–230, 242, 251
  additive 99
  chemical 4, 85, 96, 101, 107, 108, 111–113, 115, 116, 119, 120, 127, 203, 206, 209, 216, 228, 230
  classical chain 100
  emulsion 17, 124
  microemulsion 177
  surface-confined 113
  vapor-phase 91
  water-phase precipitation 38
polymerization techniques 111, 112, 114, 116, 118, 120, 122, 124, 126, 128, 130
polymers 1–8, 15, 44–47, 71–74, 76–80, 82–92, 94–96, 98–100, 111, 112, 118–120, 177, 178, 197–202, 215, 244, 247–249
  aromatic 74, 83, 99, 107
  biodegradable 53, 91, 137
  flexible 91, 104
  half-oxidized 175
  nonconjugated 44, 99
  photonic 5, 244, 247
  regioirregular 94, 95
  regioregular 95

self-doping 177
spinnable 128, 216
polymethylthiophene 99
polyolefine 5
polypeptides 5
polypyrrole (PPy) 44, 83, 87, 118, 201, 215
polypyrrole nanofibers 223, 225, 227
polypyrrole nanowires 202
polypyrrole powder 207
polysaccharides 58
polysiloxanes 103
polystyrene (PS) 5, 49, 91, 222
polystyrenesulfonic acid 124
polyurethane 51, 55, 229, 231–233, 235, 237
PPy, *see* polypyrrole
PPy composites 90, 230, 233–235
proteins 5
proton doping 111, 175, 176
proton exchange membrane fuel cell (PEMFC) 3
PS, *see* polystyrene
PVA, *see* poly(vinyl alcohol)
PVAc, *see* poly(vinyl acetate)
PVA nanocomposite fibers 57
PVA polymer powder 56
PVB, *see* poly(vinyl butyral)
PVC, *see* poly(vinyl chloride)
PVDF, *see* poly(vinylidene fluoride)
pyrrole 89, 120–122, 124, 125, 202–204, 225, 227, 228, 234, 238
polymerization of 120, 203, 227–229

saturated calomel electrode (SCE) 96, 122, 198
scanning electron microscopy (SEM) 40, 56, 138, 143, 168, 198, 219
SCE, *see* saturated calomel electrode
self-doping 177, 182, 187
SEM, *see* scanning electron microscopy
semiconductors 71–73, 77, 78, 164, 244
inorganic 77, 164
SERS, *see* surface-enhanced Raman scattering
SF, *see* silk fibroin
silk fibroin (SF) 137
silver 54, 55, 65, 227
silver ions 65, 227
supercapacitors 3, 91, 106, 107, 198, 208, 210
carbon-based 208
surface-enhanced Raman scattering (SERS) 65

Taylor cone 10, 23, 33, 35, 37, 38, 42, 131, 231
TCPs, *see* tissue culture plates
TEM, *see* transmission electron microscopy
tetraethylammonium tetrafluoroborate 206, 207
TGA, *see* thermal gravimetric analysis
thermal gravimetric analysis (TGA) 40, 230
thermochromism 93, 95
thiophene monomers 96
thiophene polymerization 98
TiO2 nanofibers 63, 64
tissue culture plates (TCPs) 146
titanium tetraisopropoxide 63
toluene 14, 18, 48, 50, 51, 97, 179, 217, 218, 223, 227
toluene sulfonic acid (TSA) 217

transmission electron microscopy (TEM) 37, 41, 55
TSA, *see* toluene sulfonic acid

ULPA, *see* ultra-low particulate air
ultrafine fibers 5, 8, 241
ultrathin fibers 2, 44

VGCF, *see* vapor-grown carbon fibers

Warburg constant 165
Warburg impedance 165–167
water-soluble polymers 15
wound dressings 46, 53, 58, 92
wound healing 2, 11, 49, 53
wounds 137

X-ray diffraction (XRD) 55
XRD, *see* X-ray diffraction

Young's modulus 146, 153, 190, 191